Cocktails

with the

Admiral

• • •

DRINKS, ESPIONAGE, AND
THE HISTORY OF THE
AMERICAN CENTURY

Other Titles by

VIC SOCOTRA

SOCOTRA HOUSE PUBLICATIONS

Cocktails

with the

Admiral

• • •

DRINKS, ESPIONAGE, AND
THE HISTORY OF THE
AMERICAN CENTURY

VIC SOCOTRA

Socotra House Publications LLC

Arlington, VA

For information about this title or to order other books and/or electronic media, contact the publisher:

Socotra House Publications LLC
4501 Arlington Blvd
Arlington, VA 22203

Cover and interior design by The Book Cover Whisperer:
OpenBookDesign.biz

Illustrations Copyright © 2024 by Vic Socotra

979-8-9872643-4-8 Paperback
979-8-9872643-3-1 Hardcover
979-8-9872643-5-5 eBook
979-8-9872643-6-2 Audiobook

First Printing 2016

Printed in the United States of America

SECOND EDITION

CONTENTS

PART ONE:
WAR IN THE PACIFIC

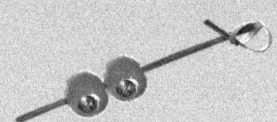

INTRODUCTION

• • •

R ADM Donald McCollister Showers started life as a farm boy in Depression-era Iowa. He once explained how his family survived the bank closures that locked up all the ready cash. His family traded farm animals for house calls by the Doctor, who shared the same problem of liquidity as everyone else.

As the New Deal was changing the face of America, Mac could see the war clouds looming. After completing college (journalism and bagpipes) in 1940, he had to ask his mother for permission to join the other 90-day-wonders in Navy Officer Training in Chicago and arrived at his first duty station in Hawaii in February of 1942, when the great capital ships of the Pacific Fleet still rested on the bottom of Pearl Harbor.

Through pure chance, he was assigned to the staff of Station HYPO, the code-breakers led by CDR Joe Rochefort, and whose wizardry enabled the Navy's greatest maritime victory at Midway Atoll.

The rest of his war in the Pacific was spent on the personal staff of Chester Nimitz, and serving as Chief of Estimates at the forward headquarters, he produced the forecast of American casualties in the impending invasion of the Home Island of Japan.

His numbers helped support the decision to open the Atomic Age.

He then strode the streets of Yokosuka five days after Japan's surrender.

Despite the rapid demobilization, he decided to stay in the Navy since he had no civilian job held for him back home.

Along the way, he met all the five-star Flag Officers, dined with Marshall Tito, and chatted up the Queen of England. He rose to make himself an admiral against the backdrop of the conflict in Vietnam. Retiring from active service, he became a trouble-shooter for the Director of Central Intelligence and had a ring-side seat for Watergate and the excesses of America's Intelligence Community.

He and his lovely wife Billie raised a great family, and in the end, he became his wife's caregiver for a decade after her illness was revealed. His support to other families in dealing with the 36-hour-day of dementia helped me immensely as my Father succumbed to the same awful disease.

He was a man in full, and his story is one that encompasses the dizzying moments of the American Century. Let's take the journey with him, in his own words.

– Vic Socotra
  Arlington, VA

THE DAY AFTER

• • •

SAYING FAREWELL AT ARLINGTON NATIONAL CEMETERY.

I t was April 16th, the Day After the funeral. The pictures had been posted, the toasts were raised at the wake at Willow. The haunting sound of the highland pipes covered the retreat of hundreds of officers and Sailors from the grave site. Mac's earthly remains were given over to the soil, and to the patient presence of his beloved Sara V., better known as Billie. The Boston Bombers who hit the Marathon finish line were still on the loose.

Mac had passed late the previous year, Oct 19th, 2012, at peace and with his family all around. I was in Colorado at the time, and I got the news it speared me with regret, though I must say Mac's final transition was conducted in the way that he preferred. His heart was giving out, the cardio experts at the Virginia Medical Center could not intervene surgically due to Mac's age. He was alert and asked what his options were. The Doctors said, "Hospice," and with that, Mac made his decision and was at peace. It was his choice all the way, just as it had been all his life.

There is a lag of months between the memorial services at Arlington and the actual interment. On this crisp day in April, the Old Guard scheduled the funeral, and Mac's earthly remains were brought, by turns, to the Old Post Chapel at Fort Myer and then to Section 66, Lot 7135 on the flat plain of the lower cemetery. The Office of Naval Intelligence turned out in force, and the honor guard of enlisted sailors and officers were led by three Vice Admirals.

I took a hundred or so and posted them to social media. Pictures are not reality, of course, but the pageant, dignity and tradition displayed is clear enough from those images. I have attended far too many of these funerals and will probably attend only one more, one in which I do not anticipate a speaking part. This was one for the ages, and is going to stand in memory for all of us.

The haunting sound of "Amazing Grace" and "Scotland The Brave" were the last echoes of the official ceremony. Mac had been a piper in his youth, and this completed another circle per his detailed instructions. The Piper is a physicist in his day job, by the way, and a good man. But his day trade is just as relevant as his skill on the pipes.

But of course there were a lot of good women and men there that gray afternoon to say farewell and Godspeed. They paid tribute to a man whose like and endurance does not come along very often. He became, with the passing of so many of his generation, an icon of his age. And a very good friend to many.

The world being what it is, we did not know of the outrage at the Boston Marathon until we got to Willow after the internment to join family and friends to celebrate what will stand in my mind as the finest example of a private interment at the Nation's place of ultimate honor.

WRITTEN APPROVAL

BACK OF MAC'S FIRST NAVY ID CARD. IOWA CITY, PAY ENTRY BASE DATE 15 AUG 1940. PHOTO AT WILLOW: SOCOTRA

Mac would have been proud, I think. Now we have to turn back to the events of the world we have all made, and it will be another grim bit of business. Mac would have had something to say about it, but his cares are not now of this world.

We are on our own now. But there is more to this. A lot more. Why don't you join me for a drink and we can talk about it.

The wine was chill in the dimness of the Willow Bar. They turn the lights down at 5:15 each afternoon to encourage the enthusiasm of the regulars. The pork spring rolls from the neighborhood restaurant menu were hot.

Humidity was down on the sun-drenched streets outside. Jake was doing some business at the bar, and Mac and I were at one of the little tall cocktail tables that line the deep brown wooden divider that separates fine dining from the usual suspects in the lounge.

I was scribbling like mad, since I have everything out of order. Mac brought some documents and books to review. He had the CIA monograph on the end of the Pacific War, and the new book on the Berlin Airlift. Just what I needed, more books, but the craving to understand is an ongoing imperative, as insistent as thirst at the end of a summer business day.

"Charles Nathan" were the Christian names of Mac's father, but he was on travel someplace. His Mom, Hedwig ("Hattie") Showers came to the door, and Captain of Police Laurence N. Ham told her why he had driven her son over from the field house at Iowa State University. The Draft Act had not been passed yet, and there were some legal niceties that had to be accommodated, even though they would soon be swept away on the road to global war.

Lieutenant Ham cleared his throat. "Your son is just ten days away from his 21st birthday, Ma'am. I need to get your written authorization for him to join the Navy.

"Are you sure you want to do this?" Hattie asked with a Mother's concern. The world, or at least the rolling low hills of Iowa was at peace. The trouble in Europe was someone else's problem for the moment. Mac nodded, and she went ahead and signed her name.

With that, Lieutenant Ham was one body closer to meeting his prodigious quota list for August, and Mac smeared his thumbprint on the faded document that Mac produced from an envelope and placed on the table in front of me.

I was careful not to drip the savory dipping sauce from the spring rolls on it, or on his draft registration that he produced as a companion piece a moment later.

"My Dad was president of the Johnson County Draft Board, and when the Draft Act was passed the next month, he insisted that I sign up, even though I was already in the Navy," he said, taking a sip of his savory red beverage. "He said no son of mine is going to be accused of not doing his duty." He shook his head at the ancient remark. "I was long gone before anyone could utter a word.

DUMB LUCK

• • •

T his was at the beginning of my fifteen-year friendship with
Mac, and just at the time, we started to take on the project
of looking at his amazing life with an eye to telling the story. It had
been a strange day in Arlington, though things are strange enough
in Washington that only the addition of nuclear weapons could
really give it some pizzazz. The Nuclear Conference wrapped up
downtown at the convention center. The First Lady made a dramatic appearance in devastated Port au Prince. The talking heads
are speculating on the impact of the retirement of Associate Justice
John Paul Stevens from the high court.

Mac and I got seats halfway down the Willow's bar, being just
a bit ahead of the rush. The place is quite fashionable these days,
and maybe it is a sign the Recession is fading. It was mild in impact
around here, though it is hard to say if the people drinking exotic
vintages ever noticed it much.

I asked him about Justice Stevens, and remarkably, it turned out he was a wartime colleague in Hawaii. I asked about his involvement in the shoot-down of Fleet Admiral Isoroku Yamamoto, which featured prominently in the published biography. "OPERATION VENGEANCE," Mac said, nodding, "and it was a long-range intercept of the Admiral's personal aircraft. The mission was based on intelligence derived from decrypted Japanese communications that outlined the itinerary for the Admiral's inspection trip to several bases in the Solomon Islands on April 18, 1943. Army Air Corps P-38s operating from Henderson Field on Guadalcanal successfully shot him down."

ASSOCIATE JUSTICE JOHN PAUL STEVENS. MAC USED TO DINE WITH HIM AT LUNCH AT THE O CLUB NEAR THE PACIFIC FLEET HEADQUARTERS IN MAKALAPA CRATER.

"It was a terrible blow to the morale of the Japanese Fleet," said the Admiral, "once they had to admit it happened. They kept it secret for a while."

Justice Steven's bio notes his contributions to the effort, though Mac doesn't remember it quite that way, and he is almost the last one in a position to know. He was a Lieutenant with Stevens, who he remembers as a polished lawyer and gentleman from Chicago. They were in the same unit at Pearl during the war, and often had lunch at the bar at the club attached to the Bachelor Officer's Quarters.

Admiral Mac is approaching his 91st summer in this world and is still chugging merrily along. He broke his elbow in a fall a month or so ago, which resulted in surgery and a marked diminution of his communications since he was reduced to one-handed typing. He does not drink anymore, on orders from his oncologist, though he doesn't mind watching. He enjoys a Virgin Mary with olives and lime and horseradish or just a ginger ale. We had begun to meet to talk about the events of his distinguished and highly secret career since most of the great events may still remain classified but are hardly a threat to anyone left alive.

I picked him up in the covered circle at the front door to The Madison high-rise assisted living facility where he lives, and we drove across the street to the Willow.

"Stevens was a good man," said Mac. "Though I have to say the modern story of his involvement in the Intelligence Community is a little overblown. I know something about that since I helped establish the Foreign Intelligence Surveillance Courts later in the 1970s."

Mac went on to say that Stevens might have been recruited to be an Air Intelligence Officer rather than a cloak-and-dagger Spook like the published biography implies.

Mac looked thoughtful as he recalled, and I dodged pedestrians to try to get parking at the curb directly in front of the door to the bar.

"Admiral Forrest Sherman created the whole air intelligence program," he said. "Sherman believed that the perfect AI was a lawyer by training, and he was right. They know how to listen, take depositions, and speak succinctly in a briefing. Admiral Forrest Sherman wanted orderly legal minds to take care of his cadre wearing the Wings of Gold. That is probably why they approached

John Paul. I don't recall him being around in early 1943, but it is possible."

"It is probably why they all got out at the end of the war to go back to lawyering," I said with a laugh.

FLEET ADMIRAL YAMAMOTO

"You might be right. LT Stevens was a new guy when I was there, and I had literally months of experience. I had reported to the code-breaking unit at Pearl (Station HYPO) in February of 1942 when the ships were still on the bottom of the harbor. That unit's unclassified name was the Fleet Radio Unit Pacific (FRUPAC) and was already humming when Mac arrived in February 1942.

Mac hadn't mentioned the dumb luck of it all. I started jotting notes on bar napkins, which surround me now. Long story short, he had never been 'destined' for anything like the illustrious career he had. It was completely by chance that he was selected to go to the six-week Investigations Course at Seattle after commissioning on 12 September of 1941.

The 90-wonder course in Chicago was intended to produce Deck Officers for the growing Fleet, and it was a complete chance he did not get orders to a ship, as most did. Some of them had enough time after commissioning to arrive at their first duty stations in Hawaii to die under the Japanese bombs.

The December 7th attack happened just after he completed the investigations course; he served briefly in the Public Affairs office in Seattle. His duties included resettling Alaskan families in the Pacific Northwest, due to the threat of war. And then it began in earnest.

All the Ensigns immediately were given orders west, toward the crisis. Half were assigned to the 16th Naval District in Manila, the other half to the 14th ND in Honolulu. Dumb luck, or it could have been the alphabetic order in which they were issued.

"Showers" fell in the second tranche of orders, and he was headed for Honolulu.

The 16th Naval District was in Manila. The kids who got those orders took the train down the coast to the Sea Port of Embarkation (SPOE) at San Francisco to the Philippines, and arrived just in time to proceed directly from the docks at Manila into Japanese prison camps-the hellholes of the South China Sea.

Some of them lived. Just Dumb Luck that Mac did not have to endure the horrors.

Mac instead got orders to the 14th Naval district in Honolulu. There was no Waikiki in those days; that was just a swamp west of Downtown and the Aloha Tower and the Matson passenger terminal. He was billeted in the YMCA near Hotel Street and the Aloha Tower where the Matson Liners landed.

The day after arrival, he walked in the soft breeze amid the scent of flowers from the Y to report to the District Intelligence Officer as ordered.

The District Intelligence Office was in a hotel the Navy had requisitioned to accommodate wartime needs, and Mac was startled to note that the Assistant DIO who received him wore two pearl-handled pistols on his belt due to his perceived threat of domestic Japanese terrorists.

"The guy was a regular Cowboy," said Mac. "He got right to the point, too. He asked me how much field investigation experience I had. I told him I had successfully completed the six-week course in Seattle, but otherwise had none. The Cowboy positively fumed." Mac chuckled, remembering the interview. "It was a guy named CDR Pease, and he was the Deputy in charge of counterintelligence, among other things.

"So, the Cowboy kind of sneered at me, saying "I need men with experience, and you are worthless to me." Mac said he rocked back in his chair and the pearl handles of the pistol poked out. "He said he had a billet out at the Shipyard he had not been able to fill since he needed all the experienced help downtown. Then he said he had just found someone he didn't need."

Next morning, Mac had orders to the Shipyard, and found Station Hypo in the basement of the Administration Building, near the bustle of the recovery efforts to salvage the stricken ships that leaned crazily at the piers or had capsized at their berths in the harbor.

By noon, Mac found himself reporting to a Navy Commander named Joe Rochefort, one of the small handful of men who understood radio, codes and the Japanese language and culture. His little unit made critical breakthroughs on the Japanese JN-25 code system, but it came at a cost.

Mac was put to work immediately, experience or no. There was a war on, after all. As mid-1942 approached, FRUPAC was literally under the gun. There were periods of round-the-clock work on intercepted messages.

CHARLETON HESTON (L) AND HAL HOLBROOK AS JOE
ROCHEFORT (R) IN THE 1976 FILM "MIDWAY".

"Did Commander Rochefort work in his bathrobe, showing up for briefings at the CINCPACFLT HQ up at Makalapa Crater late and disheveled, like actor Hal Holbrook played him in the movie *Midway*?"

"We did what we had to do in The Dungeon," replied Mac. "But do not make the mistake of thinking that Joe Rochefort was anything like the crypto-mystic Holbrook made him out to be. Joe was a pro."

I kept writing on the growing pile of napkins in front of me. Mac sipped ginger ale and told me he had been working there about three months when the frantic effort climaxed with the decryption of just enough JN-25 traffic to understand the objective of the Japanese attack.

Washington thought the Japanese were headed for Alaska,

having detected the movement of a diversionary force intended attack American interest away from the actual objective. Washington had it wrong, and that was going to be the basis of animosity and jealously that would last decades.

If you had not heard, Washington hates to be wrong.

The main body of the Japanese Navy was going to strike and seize Midway Island, and establish a bastion from which they could threaten the Hawaiian Islands.

Rochefort, with Fleet Intelligence Officer Edwin Layton, convinced Admiral Nimitz that Midway Island was the real objective. In an act of serene confidence, the Admiral gambled on the ambush that resulted in the Battle of Midway, 4-7 June, 1942. In the fight, the Japanese lost four carriers and most of their skilled naval aviators.

It was the turning point of the Pacific War.

Mac eventually transferred from FRUPAC to be Eddie Layton's assistant, and to deploy forward to Admiral Nimitz' forward HQ at Guam. He worked as chief of the forward Estimates Section on what would happen with the land invasion of the Home Islands, that helped make the decision to debut the use of the (atomic) Gadgets against Japan.

Five days after the surrender, with the help of his Boss CAPT Layton, Mac made a visit to Yokosuka Naval Base, from which the Imperial Navy ships had departed to strike Pearl four years before.

He was awarded the Bronze Star for his work with the code breakers. It was all just dumb luck that he was not in the half of his class that reported direct to be prisoners of the Japanese.

LT John Paul Stevens, USNR, finished up his time at FRUPAC,

which had moved from the basement at the shipyard to the temporary building in back of the PACFLT Headquarters.

He demobilized and got on with his life as a lawyer. As it turns out, he did pretty well. Mac stayed in the Navy, of course, and joined the people at Langley after he retired, and then retired again, and is busier in the third part of his life's work than he ever was.

JUSTICE JOHN PAUL STEVENS ON THE HIGH COURT.

He is still occasionally in touch with his old wartime buddy, and told me when I dropped him off at The Madison that he would ask him how he was planning on enjoying his retirement.

I smiled and was already compiling another list of questions. This had been a stratosphere-level. Now, we could come back and get into the weeds. I said: "We need to get together again soon, if you would be willing."

"I would be happy to," said Mac.

"I have some follow-on questions, Sir, and I want to ask them one at a time, rather than a whole war in just a few drinks. Or ginger ales, as the case might be. The one I will leave you with is this: was it all just dumb luck?"

"You don't know the half of it, Vic," said Mac. "I would be happy to continue to get together. There are a couple things I would like to get straight, while there is still time." Then he turned and walked briskly into the lobby of the Madison, his posture perfect, except for the slight list from the weight of the cast on his arm.

WELCOME ABOARD

• • •

USS *Oklahoma* capsized at her berth in Pearl Harbor.
Official U.S. Navy Photo.

I looked with dismay at the plastic bag from Bed Bath and Beyond that Mac placed on the tall cocktail table. There was a book about the Berlin Airlift that I don't have time to read and the CIA monograph that I will make time for. I waved at Peter to get a glass of the loss-leader Pino Grigio.

"The CIA publication only has about fifty pages of text," said Mac. "But the exhibits in the back are reproductions of the original documents-minutes of meetings and policy memos—they are fascinating. It shows how Truman made his decisions to use The

Bomb against the Japanese. Our estimates played a key role in that process, even if we are mentioned by name."

"Yeah," I said, smiling as Peter filled the bubble-shaped wine glass to precisely one-third of its depth. "But we talked about the end last week and missed the beginning. I want to hear about 1942, when you first got to Pearl. I want to know what you did, what the job was like, what your battle rhythm was."

Mac nodded and took a sip of his Virgin Mary. "Did you know that at one point, I had four military identification cards with four different colors of eyes?" He took off his glasses and leaned forward. His blue eyes sparkled with amusement. "Blue, Gray, Brown and Hazel.

What do you think?"

"I would say blue," I said, wondering if the pork spring rolls were on the neighborhood bar bistro menu. "Though I suppose they could be hazel or gray. I have never understood the color hazel."

Sara-with-no-H wandered by, pert and trim in her black shirt and slacks, and asked if we were hungry, and we were. Her eyes were dark as coal and just as mysterious. Really a cute gal. We ordered a couple of the tapa-sized snacks. "So we have done the big stuff, the pivot points of history and all that. What was life like? What did you really do?"

Mac contemplated the question since he was usually asked about the matters of primary interest for those of us who came after, the sinking of mighty aircraft carriers and the mushroom clouds. "Well, after I was abruptly dismissed by that cowboy Officer in Charge of the Honolulu detachment of the Navy Investigation Service,

I was sent out to the Shipyard at Pearl. The Cowboy figured he could use me to fill a lingering requirement levied by the staff of Admiral Nimitz.

It was February 1942, you know. The big boats are still on the bottom of the harbor. USS *Oklahoma* still presented her keel and one massive brass propeller pointing to the puffy clouds. Nevada was just being re-floated, and came back into the harbor for dry-docking in the middle of the month. I arrived at the Admin Building at the shipyard, presented my orders in triplicate at the quarterdeck, and was eventually shown down to the basement and presented to Commander Joe Rochefort, the OIC."

USS *Nevada* ENTERING DRY DOCK, 18 FEB 1942 NEAR THE 14ᵀᴴ NAVAL DISTRICT HQ AND CIU/STATION HYPO. OFFICIAL U.S. NAVY PHOTO.

My ears perked up. "So you went to work for him?" I made a note on the napkin in front of me, nearly knocking over the wineglass, which was wedged against the plates and cloth napkins and silverware we don't use to eat the snacks. "At the Fleet Radio Unit-Pacific?"

"No," said Mac with mild irritation, as though he was talking to a slow learner. "We were known as the Combat Intelligence Unit, the CIU. No, Joe Rochefort just said "Welcome Aboard, Ensign. He knew that I didn't have any Japanese language training or expertise with the IBM ECM Mark III punch-card tabulating machines. CDR Rochefort gave me over to Jasper Holmes, and I started to make files."

"Files?"

"Yes. Ditto files. Do you know what that is?"

"Like mimeograph machines? I remember those from grade school. And the smell." I wrinkled my nose with the memory.

"Our fingers were purple at the end of the day from the fluid. I will get to how we eventually got inside the Japanese code system, but I need to explain how this all worked."

I nodded. "That would be useful," I said and winced as I bit into a spring roll that was still blazing from the hot oil of the wok in the kitchen. *Damn*, those things are tasty.

"The Japs had introduced what we called the JN–25 code system in mid-1939. It consisted of about 33,000 words, phrases, and letters and was the primary code they used to send military, as opposed to diplomatic, messages. After Pearl Harbor, the CIU focused on cracking that code, though we also had success with a lower-grade code the used for controlling their merchant ships. That is what we gave to the submarine force."

"When did the CIU get into the code?" I asked, swirling a little of the dry white wine in my mouth to assuage the burn. I noticed there were some very attractive women at the bar, concentrating exclusively on one another and I wondered if Willow was attracting the lipstick lesbian crowd. If it was, I was all in favor of it.

"CIU had some success starting the month before I got there, in January. But it was hard. That is why the files were so important. But let me explain the way it worked. OP-20-G was the staff number assigned to the Code and Signal Section of the OpNav staff at Main Navy back in Washington. It changed to the Radio Intelligence Section the month I was welcomed aboard. We were station HYPO, named for the first letter in Hawaii. Station CAST was on Corregidor Island in the Philippines, and so on. NEGAT was at the former girl's school on Nebraska Avenue in Washington. There were over seven hundred people assigned to the effort."

"Was it NEGAT that got the "East Wind Rain" message that signaled the attack?"

Mac snorted. "There never was an East Wind message. That is all nonsense. Commander Safford was the only one who testified that he got it, and that was after the war in the testimony to Congress about who was responsible for the Pearl Harbor attack."

"OK, I know the target code, and the big issues. I am interested in what it was like to work at HYPO and what you did, how you got around, what the hours were like. You know, what it was like to be at war with the Japanese looking invincible."

"Well, to do that I will have to tell you about how I made a Frankenstein sedan, and a little about Jasper Holmes, who had quite a career writing for the Saturday Evening Post."

I knew it was time to get a fresh napkin, and waved past the pretty ladies to Peter to see if I could get some more wine.

PACIFIC WINDS

• • •

WILLOW RESTAURANT AND BAR, ARLINGTON, VA.

Mac was intent about the Winds messages. I had not meant to get into it, since it was an intellectual fight of long standing for a lot of people who are no longer alive, It was always in the background during the war, part of the long battle between the Navy and the Army code-breakers, and it burst out again, when the battle for reputation and honor was re-fought in Washington before the Joint Committee of Congress when the actual fighting was finally over.

I pushed the empty plate aside that had held the spring rolls and dipping sauce. Willow has its clockwork internal mechanisms tied to the lives of the people in the towers around it, and this afternoon was no exception. Jim the Curmudgeon Consultant came in and planted himself at the corner stool. He drinks Bud, even though Tracy's vision of her place is an upscale wine bar.

He is more than a bit like Norm on the epic bar-fly classic television show Cheers. I sometimes wonder what character I resemble, and hope it is not Cliff.

He waved, and distracted, I waved back. The line-up along the bar was decidedly more diverse than I remember the one being on television, what with the business guys and elegant ladies talking earnestly to one another, but I like it.

"Captain Safford testified that he got the famous "East Wind, Rain" message and that it was intercepted at the Security Group Station at Bainbridge Island. That was the message that was supposed to direct the diplomatic stations to burn their codes, and it was war."

He crinkled his brow. "West Wind Cloudy was supposed to mean war with Great Britain. North was the Soviet Union, I think. Doesn't matter. The message was never sent, or if it was, we didn't get it. There was a chief petty officer who was going to testify that he got the message and got it to Safford, but he was never called. The historical record is not what you think it is."

"I know that from personal experience," I said. I took a sip of pinot Grigio as Willow continued its deliberate rhythm and the lights went down precisely at 5:15, even if there was plenty of daylight outside. The sudden dimness caused us to move closer, which I suppose is the point.

"See, the Combat Intelligence Unit, working from cramped spaces in the basement of the Naval District Headquarters at the shipyard was on the defensive from the day of the attack until they started to ramp up. What Joe Rochefort and Eddie Layton and Jasper Holmes did was create OPINTEL."

I nodded in agreement. I am an acolyte of the craft pf operational intelligence, indoctrinated into the cult by the legendary CDR Mikey, the best naval analyst of his generation. He taught our generation the craft that had been perfected there in Pearl in the months that Allis reeled backwards from the inexorable onslaught of the Japanese war machine.

You had to combine overhead imagery, acoustic data, sensitive and not-so-sensitive SIGINT and blend it with highly classified HUMINT o get a clue as to where the Soviets were sending their Boomers. We were pretty good at the craft, and directing our maritime patrol aircraft to count coup on them. But the basic methodology was nothing new, and came from the basement of the Admin Building at Pearl.

The quick success of that effort enabled Admiral Nimitz to meet and decisively defeat the Japanese Navy at Midway in June 1942, just six months after the crushing defeat. It was pretty remarkable, and I have never had a good explanation of how it worked.

"So what did you do? What was your job? What were your hours?" I asked. I get all the big stuff, but I wanted the texture. I have stood watch when bad shit went down, and I remember senior people breathing down my back demanding answers when there were only questions.

Mac was a bit disconcerted. He lectures about Admiral Nimitz all the time, and the big strategic issues he was part of that are old enough to not still be secret. He is self-effacing and sometimes reluctant to take credit for what he did so long ago as a supporting cast player with giants. Then he smiled.

"It is like the cars in Hawaii. You have to remember that everything was going to the war effort. Tires were precious. So were engines. Nothing new was being made for the duration. Sometimes an automobile was worth how much tread there was on the tires."

"That is the way it was in Japan," I said, remembering the 1969 Toyota Publica station wagon I painted up in squadron colors in Yokohama. "A departing shipmate sold it to me for exactly the value of the months remaining on the Japan Compulsory Insurance policy. The vehicle itself had no worth at all, since under the Status of Forces Agreement we had an exception to the Beautification Law that forced the locals to purchase a new car every three years or pay an increasingly steep tax."

"It wasn't taxes in the war, Vic. It was just that there wasn't anything available to buy. So there was a place that made Frankensteins. I got an old Ford that had decent tires but no engine. One of he mechanics spliced in an old Chevy engine, and re-worked the mounts and I was rolling. It was not much different from OPINTEL, Everything went together to make a rolling package."

Mac fished one of the two remaining colossal olives out of the bottom of his Virgin Mary.

"So what exactly did you do?" I was determined to get to the texture of what it was like as the Pacific war lurched to the tipping point.

"I did files. Files and the overlay. We came in at all hours. There were a few officers that had families on the island, and they had something like normal hours. Most of us had nothing to do except work. Some of the linguists came in early and stayed late, sometimes around the clock." He chewed the olive and contemplated the last of the three that remained, still on the toothpick.

"The watch worked on big onion-skin paper overlay that was placed over a map of the Pacific. They annotated everything that happened over the course of the day in pencil. When the attack happened, the map only went to the edge of the Hawaiian Sea Frontier, but after the war started it was the whole Pacific. We had one in the basement that was the same as the one over at the Sub Base, where Admiral Nimitz had his HQ before they moved up to Makalapa Crater."

PEARL HARBOR SHIPYARD, WITH THE 14th NAVAL DISTRICT HQ AND SUBMARINE BASE, 1942. OFFICIAL NAVY PHOTO.

Mac was looking in the direction of Jim, who was talking to Ray, the former Marine, as Jim's long-suffering wife Chanteuse Mary looked on. He wasn't looking at them, though. He was looking into a morning long ago.

"I only went with the overlay a couple times, so I would know, just in case. Jasper Holmes took it away for the 0800 briefing every morning. The HQ was on the second floor of the building, and you got up to the second floor by an exterior staircase."

"I remember the building. Didn't it become SUBPAC later?"

Mac nodded. "Eddie Layton's assistant was a linguist named Bob Hudson. He was a lieutenant and I was an ensign. I was in the intelligence office, waiting for Jasper to get done with posting the overlay and Dunbar looked at me and said I didn't have a reason to be there and I should go out in the passageway to wait. He was a pal later, but I thought he was a first class asshole then. Probably not a very good linguist, either, since he wasn't doing anything to contribute to the code-breaking. The guys that had been trained in Japanese thought they were the kings over us reservists, whether they were any good or not."

"So you were on watch in 1942, producing the overlay?"

"Well, yes, the overlay was the big product. I understand they may actually have all of them down at the Cryptologic Museum at Corey Station down in Pensacola. They disappeared for many years, if hat is where they were. They were all ULTRA classified, the best OPINTEL compendium we had, day by day." He frowned, wistfully. "I don't imagine I will ever see them again, if that is where they are."

"So aside from the daily overlay, what were people doing in the basement?"

"We were doing files. We had boxes and boxes of hem. Jasper

had two yeomen working for him. A First Class named Bill Dunbar, and a Second Class named Irving-something, a nice Jewish kid. That made up the Information Section of the CIU."

Jim was waving for reinforcement Budweiser. Ray was looking at his watch. Jon with no "h" was tugging at his bow tie. I wear them, too, though mine are all clip-ons, an affectation I use as a political statement. Jon ties his own, of course. A couple of attractive women walked by our table, looking like they were together and happy about it.

Mac warmed to the memory. "The linguists would get the raw traffic from the cryptanalysts, and they would translate as much of it as they could, leaving blanks where they couldn't fill in the meaning. Then they brought it to us. I would then take the message and underline words of significance. I looked for the address, who it was from, who it was sent to, ship names, dates, place names, every base and every command. Then I counted up the number of underlines and wrote it on the top of the message. Bill and Irving would then make that number of ditto copies and we would put a copy into each file that matched the category. Our fingers were purple from the ditto fluid at the end of the day, and we had to scrub up with some kind of evil-smelling gunk to get it off. Over those months we built an enormous cross-reference system. When the linguists couldn't make something out, we could go to the date or the place or the unit and figure things out in context."

I chewed on the end of my pen, looking at the bedraggled napkin in front of me. "I understand the Japanese had a high level of confidence in the security of the JN-25 system. Aside from daily key changes, they never altered the code groups in the internal system, right?"

The Admiral smiled. "Yep. Our system was much better, actually an *improvement* over the German Enigma machine encryption that the Brits cracked at Bletchley Park. The confidence level over time increased with the volume of Japanese messages we cross-indexed, since they used a manual methodology. We rated our analysis of the identification as "D" at first guess, and as we became more confidence, that would increase to "A" when it was a dead certain. We would then annotate the overlay and compile estimates for transmission on the Fleet Broadcast to the operating forces."

I finished my glass of wine and nodded when Sara-with-no-"h" came by and arced one of her fabulous eyebrows. "That is not much different from the way I learned the business."

"It was just OPINTEL. Brute force analysis supported by cross-indexing and a massive filing system. No one told us how to do it. We did it as a matter of vital necessity to try to win a war. Joe Rochefort had the concept, Eddie Layton masterminded the execution, and Jasper Holmes was the genial genius who made it all work."

"So, did you feel the tide beginning to turn? Jimmy Doolittle's raid on Tokyo happened in April, and the OPINTEL picture gave you a draw at Coral Sea."

"And then Jasper came up with the idea that let us figure out the target of the last big Japanese offensive at Midway. After than, we went on the attack, and we never looked back. Oklahoma and Arizona were still on the bottom of the harbor."

Arizona still is, I thought. "When I lived there, I never begrudged going to work on Sunday at Ford Island, since the ferry from Mainside took us right past her memorial."

"Her rusting hulk is still bleeding drops of fuel into Pearl's placid waters," said Mac. "She was topped up and ready to go in December of '41. She is still bleeding, just like she did when we worked in the basement of the Admin Building."

TURRET MOUNT OF THE USS *ARIZONA* WITH OIL SHEEN, 2010

UP PERISCOPE

• • •

DAWN AT REFUGE FARM. RAYS OF THE RISING SUN MIMIC THOSE OF THE
IMPERIAL JAPANESE NAVAL BATTLE FLAG. PHOTO: SOCOTRA

E ver have that sense of dislocation, like not knowing where you
are? Makes you queasy, like the aftermath of some great and
rapid tectonic shift that slides a 7-11 Store up over its parking lot
and into the middle of an intersection.

Like my pal Mike says, the top of Everest is sedimentary rock,
and as high as it is now, was once at the bottom of a primordial sea,
way below periscope depth. He and his daughter were at Refuge

Farm briefly yesterday, and having them here was as disorienting as waking up in the wrong county.

Got me big time this morning. I realized a few moments into consciousness that I was not in my bed, but in another, also mine, but someplace else. It was as if I had come to periscope depth abruptly from the vasty deep.

I crept downstairs in the darkness and decided to stay awake. I am agitated about people I love, parents and others, and scanned the messages for portents and omens.

I discovered nothing about one situation and more than I needed on another. I have to help my brother and sister to convince my Mom to give up her home and move into something smaller and more manageable where we can get Dad the help he needs.

There is no instruction manual; how the hell is this supposed to work?

LT JASPER HOMES

I sighed. I had this big scheme for this morning. I brought my notes from the long conversation at Willow with Mac down to the farm, and was going to tell you this morning about the guy who saved his life so long ago. Jasper Holmes was the guy, who if the world was sane, would be remembered as a writer of fiction.

But of course the world is not sane, and Jasper wound up tricking the inscrutable East into the biggest screw-up in the military history of the 20th Century, which is going a fair piece.

Mac knew him well, and Jasper wrote his own book about their time together. "Double Edged Secrets" was the title, and it will say everything you need to know, like Eddie Layton's memoir that Roger Pineau and John Costello had to finish. Admiral Layton was sick at the end, and held his tongue until the ULTRA secret was finally declassified.

He savaged some of the alleged heroes of the big war, but the most guilty of them were already dead taking their medals and reputations with them. That is why my friendship with Mac means so much. When we talk, they are all still alive, heroes and villains alike.

We are sworn to silence about the middle parts of our lives, and I doubt if I will last as long as Mac to be able to speak openly about the coolest parts.

That was the subject of that long discussion the other day. Six months after Pearl Harbor, Jasper Holmes got a scheme to full in some of the blanks in all the files that Mac was cross-indexing with YN1 Bill Dunbar.

Mac has told the story of the Midway Miracle maybe a thousand times, but it never ceases to amaze me. But you need to understand a little about LT Holmes, so bear with me.

Jasper was not a career intelligence officer, but like Mac, pure chance propelled him into the most vital area of intelligence, Radio Intelligence and cryptography.

He was a line officer, originally, and a bold one. In the 1930s, he was assigned to an S-Boat, one of the original main-line USN submarines. Mac told me about them. Not much of an improvement over the German *unterzee* boats of World War One.

MODEL OF AN S-CLASS DIESEL SUBMARINE LIKE JASPER'S.

Mac says the S-Class were dank and chilly craft, which exasperated a bad case of osteoarthritis, and LT Holmes found himself medically retired and on the beach in lovely Hawaii. There are worse things in life, and Jasper used his unexpected change of career to reinvent himself. He wrote, beautifully, as it turned out, and out of deference to his day job at the University of Hawaii, wrote under a pen name he dreamed up.

"Alec Hudson" was what he chose, and as you might imagine, a naval officer with a bogus identity is something that resonates closely here at Refuge Farm. He wrote compelling accounts of life in the Submarine Force. He was published in the Saturday Evening Post, among other less prestigious places, and he was read avidly by those later caught up in the terror and excitement of the war.

His best-known works of fiction are compilations of his shorter works "Up Periscope!" and "Battle Stations."

A casual reading will betray the fact that they directly inspired the young Gene Roddenbery. The charismatic Scotsman of Star Trek ("I dinna can change the laws of Fizziks, Captain!") and Jasper's Captain Jaimeson and Roddenberry's Captain Kirk are unmistakable.

Jasper was recalled to active duty in mid-1941 with the

anticipation of the outbreak of hostilities in the Pacific. His original duties were shore-based at the headquarters of the 14th Naval District. He essentially assigned himself with the tracking of merchant vessels in the Pacific. In peacetime, it is simple enough to use ship's weather reports to provide locational data, and this association brought him closer and closer to the Spooks.

The Big Surprise on December 7th was naturally a shock. Admiral Husband Kimmel and Lt. General Short were fingered for the blame, and relieved for cause. Chief Naval Personnel Chester Nimitz was tagged to command the battered Pacific Fleet over dozens of officers more senior.

Washington would not admit that it was bitter rivalries between Army and Navy, and within the Navy itself that exacerbated the blind spot that permitted the Japanese to land a knock-out blow on the Pacific Fleet.

Jasper was propelled into the maelstrom of the response in the Combat Intelligence Unit-Station HYPO. While not initially allowed access to the sensitive COMINT mission, he swiftly became integral to it and was one of the handful of people indoctrinated into the ULTRA program.

Because he had no direct cryptographic or Japanese language experience, CDR Joe Rochefort decided Jasper's experience as a submarine officer would best be harnessed in assessing various sources of intelligence to determine the strength, composition and movements of various Japanese military units. LT Holmes became chief of the Information Section of the CIU-later known as the Estimates Branch.

That is where Ensign Mac met the man who was going to save his life. He had copies of the books that Jasper wrote, both inscribed to him, but they are long gone in the constant shuffle of belongings in a Navy life.

Edward "Ned" Beach, was a Naval Academy midshipman who went on to command submarines and have a distinguished writing career ("Run Silent, Run Deep!) of his own. He recognized "Alec Hudson" as the pen name of a naval officer, and made a point of finding out who exactly he was. He wrote about Jasper many years later, saying:

> ...[he] had become an intelligence officer at Pearl Harbor and, after the attack on the Day of Infamy, had taken on himself the particular and personal dedication to see the destruction of every ship that had participated in it. During the war, from time to time, commanders of submarines would receive by messenger, without explanation, a bottle of fine whiskey. Little by little the word got around that one of the Japanese ships sunk on a recent patrol had carried special significance for someone. In this way Jasper Holmes never left out submarines. It was through him that we would receive orders to be somewhere at a certain time – and on occasion there was a bottle of booze at the end of the trail."

That is what Jasper was up to, and with Mac and his two Petty Officers, he was going to do it with complete success.

Oh, yeah, I could tell you the story about how Jasper foxed the Japanese at Midway, and got several of the bastards, but you could always just look it up.

THE BOMB PLOT

• • •

[*12*]   From: Tokyo (Toyoda)
To: Honolulu
September 24, 1941
J-19
#83

Strictly secret.
Henceforth, we would like to have you make reports concerning vessels along the following lines insofar as possible:
1. The waters (of Pearl Harbor) are to be divided roughly into five sub-areas. (We have no objections to your abreviating as much as you like.)
Area A. Waters between Ford Island and the Arsenal.
Area B. Waters adjacent to the Island south and west of Ford Island. (This area is on the opposite side of the Island from Area A.)
Area C. East Loch.
Area D. Middle Loch.
Area E. West Loch and the communicating water routes.
2. With regard to warships and aircraft carriers, we would like to have you report on those at anchor, (these are not so important) tied up at wharves, buoys and in docks. (Designate types and classes briefly. If possible we would like to have you make mention of the fact when there are two or more vessels along side the same wharf.).

ARMY 23260   Trans. 10/9/41   (S)

THE IMAGE IS A LITTLE FUZZY, BUT THIS IS THE DECRYPTED AND TRANSLATED
MESSAGE TO THE JAPANESE CONSULATE ORDERING DAILY SURVEILLANCE
OF THE AMERICAN FLEET IN PEARL HARBOR.

There have been discussions for seven decades about how we got clobbered at Pearl Harbor. Mac did not arrive there until February of 1942, so he could not comment authoritatively on what was true or not about the claim that Washington short-sheeted the Pacific Fleet Commander on decrypted Japanese messages tracking the mooring locations of the American capital ships in the harbor.

Mac talked to Joe Rochefort about it when memories were still fresh and victory was far from certain. He said Joe wasn't sure that knowing about the message stream would have changed anything in particular, but I was reading about it again. I had time-we got snow last night, most of this winter, and Washington is Predictably Paralyzed. Willow is closed for an unprecedented second night in a row, and I officially have cabin fever. In the meantime, the memories of the placid waters of Pearl Harbor will not leave me alone. This morning it was the Bomb Plot message.

No, it doesn't refer to some nefarious scheme or plan. "Plot" is Navy-speak for information placed on a map. Like targets you might want to bomb sometime.

That is the intercepted message that might—or might not—have provided the advance information that the Japanese were interested in the precise locations of the American fleet. It might—or might not—have resulted in a different outcome in the attack. Some say that if the fleet was at sea, and might have tried to engage the attacking Japanese and been sunk in deep water, with thousands more casualties, and the proud ships lost forever, not salvaged as they were.

I got a note a week or so ago from a documentary filmmaker in the UK who is married to a pal who is a journalist based in the UK. Vicki Barker is the voice I trust on the CBS radio network. Her husband is interested in doing a documentary on the appalling way that Admiral Husband Kimmel, commander of the Pacific Fleet, and Lt. General Walter Short were hung out to dry in the wake of the disastrous attack on Pearl Harbor.

As I have told you (over and over) I lived and worked in the buildings that survived the Day of Infamy, and I was fortunate to have been a drinking buddy and tipsy Boswell to RADM Donald "Mac" Showers, the last of the JN-25 code-breakers at Station HYPO in Pearl.

My stories are sincere, but were never intended to be documentary history. I always liked to get to the loopy aspects of history-like, what was the party like at the quarters of chief

code-breaker Joe Rochefort after the amazing results of the battle of Midway became known? (Answer: *It was a good one*).

Elliot Carlson is a professional historian who was working on his extraordinary biography of Rochefort through the same period, and detailed the scandalous story of Joe's removal from command of HYPO by ankle-biting careerist hacks in the moment of his greatest triumph.

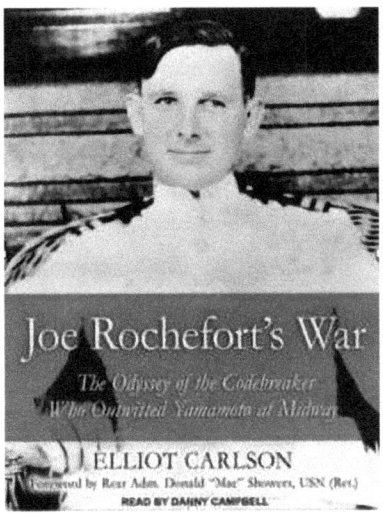

There is still a lot of emotion in all this. I know a pompous jackass here in town who was going around trying to discredit the story of how Joe's band of Japanese linguists and cryptologists identified the target of the attack on Midway.

As you may recall, ENS Mac Showers was sitting at his desk in the Dungeon when Jasper Holmes outlined his scheme to tell the Commander at Midway (by way of secure submarine cable) to report by unencrypted radio that the freshwater plant on the coral atoll was malfunctioning. Holmes said that when the Japanese copied the radio transmission, it might reveal the identity of the target for the coming attack, and permit decisive action by the American Fleet.

Joe Rochefort liked the idea, and it worked. Within a few days, an encrypted Japanese message was de-coded that revealed the target—"AF"—was having fresh water problems.

It wasn't perfect intelligence, but that is about as good as it gets in the world of COMINT. Joe was wary of the testy relations he had with the radio intelligence people back in Washington. He didn't want the conclusion of the elegant subterfuge to be compromised by suspicion of who figured it out. So he provided the information to the Fleet Radio Unit in Melbourne, Australia, (FRUMEL) to have *them* report it.

Joe had a sign behind his desk that read: "There is no limit to what **you** can do, so long as **you don't care who gets** the **credit**." And so, sixty years after the fight, people were still fighting about stealing credit for what Jasper and Joe did in the basement of the 14th Naval District.

With things like that going on, Elliot relied heavily on Mac's razor-sharp recollections to keep things straight. We are still attempting to get Mac's 26-hour oral history transcribed and declassified—money and personnel bandwidth has been a challenge but we continue to press.

Elliot's interviews with Mac were conducted in a much more focused manner than the ones I was doing, but they were complementary in nature.

I have asserted that the real villain in all this was Admiral Richmond Kelly "Terrible" Turner, who had been Chief of War Plans at Main Navy before the war had been instrumental in denying Hawaii access to the high-level Japanese Diplomatic messages in the Purple intelligence stream. That, and the famous Bomb Plot message are described In "Joe Rochefort's War."

The Bomb Plot message is presented on page 154 and onward in the book. The question from the filmmaker was about exactly that. I wrote him back that *"Arlington Hall Station (across from where I now live) was where the Army conducted its code-breaking. Their charter—and part of the division of labor with Navy—was to attack the Japanese high-level diplomatic code (Purple) as well as less important cypher systems used by the Japanese diplomatic corps."*

HYPO's mission was only to track Japanese Naval ships. The Bomb Plot message was in a different cypher system than Purple, known as "J-19," and was considered a secondary priority to breaking Purple traffic. The message had originally been copied on 24 September at the Army intercept site at Fort Shafter on Oahu. They lacked translators and machine support at Monitoring Station-5 (MS-5), and so the cable in question was shipped back to Washington for processing.

According to Elliot, the message was eventually deciphered a few weeks later and circulated to the Army and Navy Radio Intelligence brass. They determined the request for detailed anchorage positions was routine and the message was then filed away without action.

It did not come to light until the Pearl Harbor inquiries were held—the first in the wake of the attack, and the second, more elaborate one, after the war was over and the sense of urgency slightly less.

Later, when the existence of the messages were revealed, Rochefort himself was not confidant that he would have recognized the Bomb Plot message as significant. He remarked that he might have just chalked it up to the obsessive nature of Japanese collection philosophy.

That was emphatically not true for Admiral Husband Kimmel and Lt. Gen. Short were, since they had already been railroaded and scapegoated. Both were back in Washington before the existence of the Bomb Plot message was revealed to either. Eddie Layton would probably have been the one to bring it to Kimmel as his Fleet Intelligence Officer, but it never made it back to Hawaii.

Nor did the high level Japanese diplomatic traffic transmitted in the Purple system. Richmond Kelly Turner made sure those messages did not get to Admiral Kimmel—including the ones that directed Japanese diplomatic staff to have things wrapped up by the end of November 1941. The Kido Butai, the IJN Main Body, sortied for Pearl on the 26th of that month.

At the dawn of the age of machine encryption, special machines had been constructed to assist the laborious process of breaking the contents of the messages—there are some great stories to tell about black-bag jobs and the like conducted by Naval Intelligence to enable the penetration of the codes. Station CAST in the Philippines had one. Station NEGAT in Washington had one.

The one intended for Pearl was in Bletchley Park in the UK. There is a story about that, too, and maybe we will get back to Beach Gradients one of these mornings soon.

But if anyone tells you that Joe Rochefort was wrong, or that the guys in Australia had anything to do with Jasper Holmes and his great idea, just tell them they are pompous partisan windbags.

Oh, and it is good to have friends. Thanks to that, I can give you a look at two of the Great Americans who labored in The Dungeon in the basement of the 14th Naval District at Pearl Harbor:

FOXING THE SUN

• • •

OLD JAPANESE COMIC BOOK, *ARMY OF MICKEY'S FOLLOWING THE BATTLE FLAG.* IMAGE COURTESY: JENNIFER

The trick to understanding how Jasper Holmes foxed the Empire of the Sun is contained in the methodology they used to create OPINTEL. It was a fusion of methodologies: an 'all-source' approach to integrate communications intercept, code breaking, photographic, open source information, topographic, hydrographic and human intelligence, all cooked up in the processing unit of the Combat Intelligence Unit in the basement of Building One at the Shipyard at Pearl.

Of course, it started literally the day of the attack in December 1941. Commander Joe Rochefort was Officer in Charge (OIC) of the

Radio Intelligence Unit (Station HYPO). The riddle was wrapped in an enigma as is common. The whole thing was covered by the euphemism of the CIU.

Mac told me at the Willow bar that the Navy cryptanalysts were not able to read operational Japanese Navy message traffic-the JN-25 code system at the beginning. Jasper Holmes had made some progress tracking merchant ships, the information on which was available through Lloyds before the war swept all the commercial activity into the *levee en mass* war efforts of the respective belligerents.

Mac joined Holmes and his two yeomen in February, by chance, and some significant progress had been made, though trial and error. Joe Rochefort was to devote all his efforts to breaking the JN-25 code, and enlisting the support of the growing mountain of cross-indexed messages to glean the true meaning of each five-digit code group within them.

Once enough sense was made out of an intercept, it was turned into actionable information to support the pencil annotations on the daily onion-skin overlay that went to Admiral Nimitz each morning, and to transmit to the Fleet.

"That is pretty crazy, Admiral. A stream of coded messages going out from the Japanese to their Fleet being copied by us, decrypted to the degree you could, and then re-encrypted and sent out to U.S. units."

"Our system was pretty good," said Mac, as I was trying to catch Peter's eye and get the tab so we could take Willow's moveable feast on the road home. I had dozens of napkins in front of me, all covered with cryptic pen strokes. "We called our machine code system the

ECM-literally the Electric Coding Machine. As far as we could tell, it was unbreakable."

"That is a useful thing in a code system," I said. "When we found out the Walker assholes had been feeding the Soviets our codes for years, and that the machines themselves had been compromised by the North Koreans and Vietnamese, we realize how badly we were screwed."

TRAITOR JOHN WALKER

"It is just good that we did not have to go to war with the Russians. But John Walker started supplying code material to them in 1967, and there is evidence that the Pueblo incident was a result of them wanting the machine enciphering equipment to run the keying material. The North Koreans were happy to help."

"Everyone is upset about Bradley Manning and the Wiki-leaks thing, but Walker compromised millions of naval messages."

"Yep," said Mac, looking a little wistfully at the remains of him Virgin Mary. "And some good people died and many bad people didn't because of those traitors. Communications security is how you save lives and win wars."

Eventually, I got home and thought I might swim, and placed the stack of napkins next to the computer to try to work my own style of decryption in the morning. I was about to head down to the pool before Joanna the Polish Lifeguard locked up for the night when my pal the attorney called me up with a question. He was still at the office, on San Diego time.

"I have been reading about your conversations, and you know I have a legal mind. So, I was wondering what our Morse code operators actually copied down when listening on their intercept radios? It just occurred to me that Japanese is an ideograph language, so instead of words, even in plain text, they would have to have a series of cipher groups to stand for individual ideographs."

ENIGMA CODE MACHINE. PHOTO: NATIONAL CRYPTOLOGIC MUSEUM

"I don't know," I said. "I am much more familiar with the attacks on the German Enigma machines and the algorithms based on machine encryption. I am a liberal arts guy, but my understanding of that process is that when the rotors were set to the daily key setting, the operator would type in words in plain text and the machine would scramble it up into hundreds of thousands of variations. It would arrive at the distant end and be processed by a machine with the same rotor settings and come out in the original text."

My attorney was not satisfied. "So did the COMINT operators copy something like would look like a string?

He rattled off a string of numbers as an example: "59442 break 66999 break 30522 break 99911 00456. That would produce an ideograph cipher-group that meant "Proceed full speed Point A"?"

"I don't know," I said. "I will have to ask Mac." I made a note to call him on Sunday. The Attorney has a reason to take this seriously. He had spent a year in Vietnam, right on the Cambodian Border, and walked into a Soviet COMINT site that had been located on one of the islands next to Vietnamese waters, eavesdropping on U.S. battlefield communications.

The North Korean capture of Pueblo and the treasonous Walker disclosures ensured that they got great information.

By Sunday, and I was already depressed with the prospect of starting another working week. The pool had been unusually chilly when I swam, and the clouds had gathered in the afternoon and even this early in August there was the hint of the Fall to come.

I realized I was going to have to start planning shortly for an exercise activity that did not include the Big Pink pool.

To get my mind off the gloom of Black Sunday, I called up Mac at his apartment at the Madison where he lives. He was catching up on correspondence anyway and happy to talk. After some pleasantries, and making a tentative date for our next trip to Willow, I tried to paraphrase the attorney's question.

"So, were there actually three steps in the decryption process of the JN-25 Naval Code. First to copy it accurately, then break the numerical codes that stood for Japanese ideographs, then to translate those ideographs into English?"

Mac hesitated briefly. "I'm not qualified to give a complete answer, Vic. I was neither a cryptologist nor a linguist. I was a Deck Officer with the basic investigations course as professional background. But here is how it worked in the basement of Building One. Our intercept operators copied JN-25 five-digit code groups, as your friend guessed."

There was a pause on the line as Mac recalled. "These were not actual code groups but were five-digit groups taken from the code dictionary of 40,000 to 50,000 groups. Then, to encipher the message, the Japanese communications clerk would go to an Additive Table of random five-digit groups that were added to the basic code group, with no carrying to the adjacent digit."

"Then, to decipher the text, the receiving clerk would take the incoming encrypted message and go to the same table and subtract the additives to expose the clear text of the message, which can then be read from the code groups in the code dictionary.

"That doesn't sound so hard," I said.

"That was a very simplistic description of what the Japanese operators had to do, Mac snorted. "When we got the five-digit groups of the message through the intercept operators, the processing unit where I worked began to attack it. Our first task was to determine the additive numbers that were used and subtract them from the transmitted text. We called that step "stripping the additives," and it got us to the basic code group. Then we would apply cryptographic skills to determine the possible meaning of each code group, one by painful one. Over time, with luck, some good guesses, and maybe an occasional crypto error by a Japanese clerk, the linguists could begin to recover the meaning of each individual code group.

"After months of effort, we could begin to read fragments of messages, and then things became easier. Greater volume of traffic, a few more operator errors, and greater familiarity with the system all contributed toward being able to read messages in greater detail. Of course, then the Japanese would change the additive tables, and we'd have to start all over again. The one thing that made all the difference was that the basic code group dictionary wasn't changed during the entire war."

I thought about that for a moment. "So when Jasper Holmes thought up his little plan to make the Japanese tell us the point of attack in their last major offensive, there was a missing code group for Midway island?"

Mac nodded. "Yep. Washington was convinced that the Japanese were going to land at Dutch Harbor in Alaska. Jasper thought the communications activity that suggested a northern attack was a ruse, and that the main battle force would try to take Midway. So, he convinced Rochefort and Eddie Layton to send a message via the secure underwater cable get the naval station on the island to broadcast that they had problems with their water evaporators."

"Which was then copied by the Japanese, who reported the five-digit code group of the place of attack had water problems."

"Correct. It happened right by my desk in The Dungeon. Then all Eddie Layton had to do was convince Nimitz to gamble everything he had on the chance that Jasper was right."

"What a story," I said, looking at the increasingly cloudy sky.

JAPANESE HEAVY CRUISER OF THE MOGAME CLASS ON FIRE AFTER
ATTACK BY PLANES OF TASK FORCE 16, 6 JUNE, 1942.

Mac cleared his throat. He quit smoking years and years ago, but he enjoyed his Lucky Strikes during the war. "I am not a linguist or a cryppie, so I cannot give you a more technical description of the process, and all the persons I worked with during WWII who could do so are six feet under. Someone at NSA might be able to do so, but you'd probably have to be cleared into his domain before he'd tell you, and then he might have to shoot you afterward. The Fort is kinda funny about such things, you know."

I thought about the Walkers, and a war that was won, and one that was lost, and thanked Mac for his time. He may have told the story about Jasper Holmes a thousand times, but I will never lose my amazement for it.

"Next time at Willow we need to talk about how things changed after Midway at the Combat Intelligence Unit."

"OK," said Mac. "I'll tell you how it worked at the Joint Intelligence Center, Pacific Ocean Area when we went on the offensive."

"Cool," I said.

Meanwhile, this episode with Mac at Willow is one of the crucial turning points in the affairs of the Nation and of the people who serve it. Or profess to do so, anyway. This is one of the central issues in the history of the American Intelligence Community, and about why Mac felt so strongly about getting Joe Rochefort the recognition he deserved, and for which Mac and Jasper Holmes and so many others tried to rectify. In the end, Mac succeeded, but the wars within the secret world continue. In view of all that has happened lately, it would appear that we are still in for a wild ride, and some of the echoes of long ago events continue to reverberate in our own times.

WITH THE GOLDEN PELICANS AT MIDWAY

• • •

PBY crew who discovered Japanese fleet, Battle of Midway.

W hen I was interviewing Mac, it was seventy odd years from the exact time the Golden Pelicans of VP-44 were manning up their PBY *Catalina* and going out to see if Eddie Layton's bold prediction about the location of the main Japanese battleforce headed for the invasion of Midway Island was going to be true.

I am going to be headed out to Pearl next week, just missing the ceremonies, for a perfectly good reason.

My 92-year-old drinking buddy is one of the last survivors, and considering the significance of the 70th anniversary of the great battle, and the fact that so few of the participants are left, Mac is going to take his family out there. My boss Jake has been asked to make a speech, along with Mikey, who was the Big DNI for a few years.

I worked for a couple times in his amazing career. I am going to tag along to pay tribute and see once again the Islands where my sons were born.

Jake had an interesting story. He made one-star while on the USCINCPAC Staff, and was the "duty Admiral" one weekend when a special flight was laid on to scatter the ashes of ENS George Gay, the "sole survivor of Torpedo EIGHT," from the CINC's dedicated P-3 Orion aircraft on the waves where the battle was fought so long ago. After he was shot down with the rest of his squadron, he spent an eventful day amid the Japanese fleet. It was Gay's colorful eye-witness account of the sinking of three IJN Carriers that electrified the nation on the cover of Life Magazine.

Of course, it wasn't completely true. Gay's story was useful in that it provided the Combat Intelligence Unit an alternative explanation to why Admiral Nimitz knew so much about the devastating results of the battle. George had a story that did not include the vulnerability of the JN-25 coding system used by the Japanese.

I met ENS Gay at one of the big fly-ins at Oshkosh years ago, and he signed a copy of his book about the battle, "Sole Survivor." He actually made LCDR before the war was over, and despite a follow-on career flying for Trans World Airways, he would always be "ENS Gay," the only one of thirty squadron mates to survive the battle.

Jake noticed two older gentlemen who were on the CINC's plane. For the life of him, he could not recall their names, but the event of the day were interesting. A stone was dedicated at Midway, and a harrowing event followed on a long flight 700 miles to the west as an aircrewman attempted to deposit ENS Gay out the hatch. He asked me to check it out—he thought the old guys might have been the pilot and co-pilot of the PBY whose contact with the Japanese provided plausible cover for the JN-25 intercepts that enabled Mac and station HYPO to offer Chester Nimitz the chance to bushwhack the Japanese.

The flight Jake was on covered the thousand nautical miles to the northwest of O'ahu. Midway, outer rampart of the Hawaiian Island chain, had been recognized to be strategic as early as 1867 when the Secretary of the Navy, Gideon Welles, directed that Brooks Island, as it was then known, be claimed and surveyed for the United States. In 1869, Congress appropriated $50,000 for dredging an entrance channel to the lagoon. Over time, the atoll became a relay station for the trans-Pacific cable that linked the American colony of the Philippines to the Mainland.

Pan Am used the lagoon for a refueling station, and created a modest R&R and hotel facility in Mid-Pac. Pan Am was a great institution. I am sorry it is gone, though we saw remnants of their Pacific operations all over. I bought a copy of "Shattered Sword," the latest in the long series of the history of the battle by Parshall and Tully, using extensive Japanese sources.

I personally found "Joe Rochefort's War" by Elliott Carlson published later to be much more approachable and human scale.

Anyway, there are remarks to be drafted and stories to be told, so I thought I would take a whirl through the fight and get oriented so I could ask some pertinent questions to Mac, who was still alive and sitting at the Willow Bar, since he lived the experience.

By 1942, Midway had become the front line of the expanding area of the Japanese-controlled Pacific Ocean. Wake island, the next of the strong of pearls across the broad blue waves, was a thousand nautical miles to the WSW. It was occupied by the Imperial Army in the opening days of the Pacific conflict, the Navy and Marine Garrison fighting hard before surrendering to overwhelming force.

The American spine had stiffened in the months of reeling defeat since the attack on Pearl—and the work of Station Hypo contributed to the ability to put maximum fire-power at the point of attack. The fight at Coral Sea was inconclusive, but shored up the route of the vital re-supply lines to Australia.

In May 1942, Admiral Isoruku Yamamoto, commander in chief of the Japanese Combined Fleet, was prepared to shift his offensive operations north and east from Coral Sea and knock out the USN once and for all.

His staff devised an intricate plan called Operation MI, to draw out the U.S. Pacific Fleet by attacking Midway. Using Midway as bait and gathering a vast naval armada of eight aircraft carriers, 11 battleships, 23 cruisers, 65 destroyers and several hundred fighters, bombers and torpedo planes, Yamamoto planned to crush the Pacific Fleet once and for all.

He would ambush the mauled American Fleet at Midway Island, and then secure a base for land-based aviation to regularly strike Pearl Harbor.

Alerted by his code-breakers that the Japanese planned to seize Midway, Admiral Chester

W. Nimitz, commander in chief, Pacific Command, flew to the atoll on May 2, 1942, to make a personal inspection. He wanted to size up the defensive posture and the character of his Navy and Marine commanders on scene. Commander Cyril T. Simard, USN, was the air component commander, and Lt. Col. Harold D. Shannon was the Marine ground commander.

They gave the Admiral the grand tour.

Following his personal inspection, Nimitz took Simard and Shannon aside and asked them what they needed to defend Midway. They told him their requirements.

"If I get you all these things, can you hold Midway against a major amphibious assault?" Nimitz asked the two officers. "Yes, sir!" Shannon replied.

It was good enough for Nimitz, who returned to Oahu. On May 20, Shannon and Simard received a letter from Admiral Nimitz, praising their fine work and promoting them to captain and full colonel, respectively. Then Nimitz informed them that the Japanese were planning to attack Midway on May 28; he outlined the Japanese strategy and promised all possible aid.

There were issues in the massive build-up. On May 22, a sailor accidentally set off a demolition charge under Midway's fuel farm. The explosion destroyed 400,000 gallons of aviation fuel, and also damaged the distribution system, forcing the defenders to refuel planes by hand from 55-gallon drums.

All the while, the Marines continued digging gun emplacements, laying sandbags, and preparing shelters on both islands.

Barbed wire sprouted along Midway's coral beaches. Shannon believed that it would stop the Japanese as it had stopped the Germans in World War I. He ordered so much strung that one Marine exclaimed: "Barbed wire, barbed wire! Cripes, the old man thinks we can stop planes with barbed wire!" The defenders also had a large supply of blasting gelatin, which was used to make anti-boat mines and booby traps.

On May 25, while the work continued, Shannon and Simard got some good news, courtesy of Joe Rochefort's band of code-breakers at Station Hypo. The Japanese attack would come between June 3 and 5, giving them another week to prepare.

That same day, the light cruiser St. Louis arrived, to deliver an eight-gun, 37mm anti-aircraft battery from the Marine 3rd Defense Battalion and two rifle companies from the 2nd Raider Battalion. Beginning on May 30, Midway's planes began searching for the Japanese. Twenty-two PBYs from Lt. Cmdr. Robert Brixner's Patrol Squadron 44 (VP-44) and Commander Massie Hughes' VP-23 took off from Midway lagoon, then headed out in an arc stretching 700 miles from Midway in search of the main body of the invasion force.

By June 1, both Sand and Eastern islands were ringed with coastal defenses. Six 5-inch guns, 22 3-inch guns and four old Navy 7-inch guns were placed along the coasts of both islands for use as anti-aircraft and anti-boat guns. As many as 1,500 mines and booby traps were laid underwater and along the beaches. Ammunition dumps were placed all around the islands, along with caches of food for pockets of resistance and an emergency supply of 250 55-gallon gasoline drums.

MIDWAY ISLAND-ACTUALLY THREE SMALL ISLANDS IN A MIDDLING-LARGE LAGOON. NOW ADMINISTERED BY THE FISH AND WILDLIFE SERVICE, THE ISLAND WAS A KEY MID-PACIFIC NAVAL BASTION FOR A CENTURY.

Six new Grumman TBF torpedo bombers arrived on the island that day, commanded by Lieutenant Langdon K. Fieberling. None of the TBF pilots had ever been in combat, and only a few had ever flown out of sight of land before. The TBF would later be named *Avenger* in honor of its combat introduction at Midway

Midway had practically everything it needed for its defense. Along with the 121 aircraft crowding Eastern Island's runways, Midway had 11 PT-boats in the lagoon to assist the ground forces with anti-aircraft fire. A yacht and four converted tuna boats stood by for rescue operations, and 19 submarines guarded Midway's approaches.

On June 3rd, a VP-44 Consolidated PBY commanded by ENS. "Jack" Reid, was assigned sector search west by southwest, which was in the general area for a possible encounter with the IJN twin-engine *"Betty"* bombers that flew out of captured Wake Island. The crew hoped for an encounter with one of the Japanese aircraft. The night before one of the crew members had traded some beer for 5 new explosive .50 caliber shells from a B-17 crew. The ordnancemen on the crew had loaded them on the port waist gun.

CAPT JACK REID, USN-RET

The flight came to the end of their outbound 600-mile leg with no sightings. The crew urged Jack Reid to go further to see if they couldn't make contact with a *"Betty."*. Jack checked with navigator Bob Swan and was assured that they still had plenty of fuel to go another 20 or 30 minutes on the present course. Jack agreed to the plan and told Bob," just give me as heading when we get to the end of the time limit."

The flight continued on for the allotted time and as Bob was about to give Jack the new heading for the dogleg and at that instant Jack spotted specks on the horizon. He gave the binoculars to the second pilot Gerald Hardeman saying: *"Are those ships? I think we've hit the jackpot."*

Hardemen concurred. Moments later, John Gammell, in the nose turret, sang out, "Ships dead ahead, about 30 miles dead ahead." a radio message was immediately sent to Pearl Harbor saying, "Sighted main body." Minutes later, a second message, *"Bearing 262, distance 700 miles."*

Nimitz' headquarters at Pearl Harbor and Fletcher's carriers also received Reid's "Main Body" message. Since they expected Nagumo to be coming from the northwest, not west/southwest, this message briefly posed a problem. But Nimitz stuck with his intelligence forecast, and radioed back to the carriers *"The force sighted is not, repeat not, the Main Body."*

That is the essence of this story, the necessity of covering the vulnerability of the codes with plausible deniability.

Jack Reid scouted the force for another two hours, not knowing which part of the elephant of the huge Japanese formation he was observing. He kept the Catalina at low altitudes and came up from different positions, counting the sightings at each one and radioing the results. The long wakes in the ocean from the armada led him to either port or starboard of the ships. He knew full well, if detected, they would be hit by swarming Zero-sen fighters.

The force Jack's crew had sighted consisted of 17 ships, battleships, cruisers, destroyers, and transports headed for Midway. It was not the target that Joe Rochefort predicted carried the largest threat to the island—the fast carriers.

Commander Robert A. Swan, of Santa Rosa, California, was

the navigator on 44-P-4. He always had a familiar smile, with a personality to match. A Naval Reserve PBY pilot in 1942, Bob was the navigator on Jack Reid's Catalina.

44-P-4 landed back at Midway with little fuel to spare, and one of the two massive engines sputtered out after they landed in the lagoon. When asked why they were able to stay aloft for an additional 3 hours, Bob replied, "Raymond Derouin (the plane captain) has three dependents-a wife and two daughters. He always puts in an extra 50 gallons for each one."

Bob continued in patrol aircraft for the remainder of the war and stayed in the Naval Reserve after its conclusion, retiring as a commander.

VP-44's greatest contribution to victory had been made, but the battle was only now being joined. On June 4, Reid and Hardeman flew more than 14 hours, again providing important contact reports. Indeed, he had become an important set of "eyes" for the U.S. Fleet. His PBY was attacked by Zeros and by AAA on a Japanese cruiser, but Jack got his aircraft and crew to safety up in the clouds. Later he landed in the lagoon at Midway, and as he taxied toward Sand Island, one of his engines sputtered out for lack of fuel. Nonetheless, he was up and flying the next day, searching the Pacific for lost pilots and crews.

Jack Reid stayed in the Navy after the war and retired with more than 30 years of service as a Captain, setting up his home in Aptos, CA.

The last members of 44-P-4 have passed on. Jake got a chance to meet them, and shoot the shit in the plane on the long flight west.

I wish I had the chance, but you have to take what you can get. I still think about ENS Gay, and his card table at the Oshkosh Air Show, and the way he returned to the Battle of Midway from the back of the CINC's plane.

BATTLE OF MIDWAY NATIONAL MEMORIAL ON THE ATOLL.

On September 13, 2000, Secretary of the Interior Bruce Babbitt designated the lands and waters of Midway Atoll National Wildlife Refuge as the Battle of Midway National Memorial, "so that the heroic courage and sacrifice of those who fought against overwhelming odds to win an incredible victory will never be forgotten."

The monument reads, in part: "They had no right to win. Yet they did, and in doing so, changed the course of a war."

This is the first National Memorial to be designated on a National Wildlife Refuge. Numerous historic sites portraying man's history on the islands since the early 1900's are protected by the Fish and Wildlife Service, including several World War II defensive positions that were designated a National Historic Landmark in 1986.

The murals from the old Base Theater are on loan to the Pacific Aviation Museum on Ford Island, and they are spectacular. And now you don't have to go to a wildlife refuge to see them.

NONE DARE CALL IT ...

• • •

Chicago Tribune

"JAP FLEET SMASHED BY U.S. 2 CARRIERS SUNK AT
MIDWAY NAVY HAD WORD OF JAP PLAN TO STRIKE AT
SEA KNEW DUTCH HARBOR WAS A FEINT"

– *Chicago Tribune* Headline, Sunday 7 June 1942 that disclosed ULTRA information and may have tipped the
Japanese that they had some real security problems.

I got stuck on something that might be treason this morning, not
the Bergdahl thing, but one that happened in the months after
the victory at Midway. I am happy the wandering Sergeant is back
in U.S. hands, and I am also expecting the Army to do the right
thing, and bring him up before a Court Marshal on Article 85 or
86 charges under the Uniform Code of Military Justice (UCMJ).
I don't really care what the verdict is. It is not the first time some
really ambiguous things have happened, and I am content with the
rough justice the military deals out. All of use who served in Korea
know about another one. U.S. Army Sergeant Robert Jenkins was
assigned to the Republic of Korea back during the Vietnam War,
decided he did not want to participate, and defected to North Korea.

Bad career move—it took him 40 years to get out. The Army tried him when he came home, slapped his wrist with 24 days in the Stockade, and gave him a bad conduct discharge, or what we knew as the Big Chicken Dinner (BCD). At least it saved the taxpayers from having to shovel out the back pay, and if Sergeant Bergdahl deserted or went AWOL, we ought not to have to pay him for it.

But it is funny that it is always the troops who wind up in these situations. I was looking at the daunting pile of manuscript associated with Mac's story this morning and realized there are a lot of officers who never paid jack-squat for anything. Some of you may recall the essence of the rest of this rant from August of 2010. Mac and I were wrapping up 1943 in discussions at Willow, but we came back to the matter of how Joe Rochefort, the gifted cryptologist, was hung out to dry by the Rear Echelon MFs.

I should stay away from ancient evil, but I need to get to it to describe the burgeoning intelligence effort that took Mac from underlining message and arranging IBM punch cards to making rubber topographic maps of remote islands that no one ever heard about.

I spent the first hour waking re-reading parts of CAPT Eddie Layton's book, "And I Was There" in preparation for today's outing, which was intended to talk about 1943. But there are some unburied dead from the period after June, 1942.

I chuckled as I read about Joe Rochefort's Deputy in the Combat Information Unit, Thomas "Tommie" Dyer. Layton said that he had the best collection of pin-ups in the Pacific under the glass on his desk. I made a note to ask Mac about how lurid they really were.

Of course, what was on the top of Dyer's desk obscured the pictures most of the time. He said later that Rochefort and the other analysts, including himself, kept most of the five-digit code groups in their heads, and the desks were covered with hundreds of partial decrypts. They worked port-and starboard watches most days, and around the clock before the Midway break that identified Admiral Yamamoto's target.

A newly arrived Yeoman once cleaned off the desk when Dyer was sleeping, and there was holy hell to pay, since like the code groups that floated in endless strings through his brain, he knew where every page that lay above the pin-ups was. The effort paid off. With Jasper Holmes trick, the target was identified, and with superhuman effort, a Lieutenant named Joe Finnegan managed to construct a table that cracked the super-encryption on the date of the attack.

Admiral Nimitz crossed the Rubicon at a major inter-service conference on the 27th of May; he believed Eddie Layton's prediction that the Japanese carrier would launch the attack "on the morning of 04 June, from the northwest on a bearing of 325 degrees."

Eddie was spot-on, though there was uncertainty up to the last moment. The Japanese had made a pre-invasion change of additives, and HYPO was in the dark on the eve of battle.

I won't attempt an account of the struggle itself, since better people have done that. In The Dungeon, Mac was placed on a desk under a bunny tube that would deliver messages by pneumatic pressure. Those quaint delivery systems were still in the fleet when I arrived decades later and the rattle of the arrival of the hollow projectile was always exciting. But only a few intercepts arrived as the titanic struggle raged.

What interests me as a Spook is what happened afterward.

CAPT JOE ROCHEFORD. OFFICIAL U.S. NAVY PHOTO.

Washington had been predicting that the attack could happen in the middle of June, and fall upon either Alaska, or perhaps to the south. Had anyone in the Pacific paid attention to their better-resourced predictions, the Japanese would have been using the Fleet Post Office code they had assigned to Midway Island.

It is said that victory has many fathers, and defeat only one.

The Redman Brothers, Joe and John, in the Office of Naval Communications and OP-20G (Radio Intelligence Section), respectively, had immediate access to the senior brass of the Navy and took credit for providing the intelligence that enabled the victory.

Anyone who has been forward and afloat knows that the Shore Establishment always wins, and the chance of victory is enhanced the closer your desk is to the flagpole at the Pentagon, or in Mac's time, at Main Navy.

Once victory was certain, historian Stephen Budiansky quotes Joe Rochefort told everyone at Station Hypo that he "didn't want to see them for three or four days." He expected everyone would just go home and catch some sleep.

Instead, a house party on Diamond Head was convened. Budiansky quotes Rochefort as saying it was a "straight out-and-out drunken brawl" that lasted the entire three days. Then everyone shook off their hangovers and went right back to twenty and twenty-two-hour shifts to tackle the new code book and additives that the enemy had introduced into JN-25 before the battle.

I need to ask Mac about that. Or retiring Associate Justice of the Supreme Court John Paul Stevens.

I am much more likely to see Mac at Willow.

But the real battle was just beginning thousands of miles east of Midway. The assertion that Washington's Station NEGAT had been right was breathtaking enough, but there was an implied task contained in taking the credit for other people's success. They had to discredit Joe Rochefort and Eddie Layton.

The coup engineered by the Redmans to oust Joe Rochefort from his post in The Dungeon is quite extraordinary.

The *Chicago Tribune* Affair reveals the banality of institutional evil, personal ambition and the power of The Green Door, what we called the gateway from reality into the secret world that went on behind it. Here is the deal: a war correspondent named Stanley Johnson was embedded with the operating forces that went to Midway. He provided the article on which the re-write man in the Windy City based the headline slugs up above.

Johnson was a classic exemplar of the knock-about, wise-cracking newshound made popular when Time Magazine was edgy journalism three-quarters of a century ago. Born in Australian, he wore a big black mustache and had served in the Australian Army in World War I. He roamed Europe and Asia for years after the war, perhaps a victim of Hemingway's version of PTSD. I guess we are not supposed to use the "D" anymore, but sometimes I mess up. Johnson wound up as a stringer for the *Tribune's* London bureau. He came to the U.S. after the fall of France and married a former showgirl he had met in Paris years before.

He became a U.S. citizen, and his free-wheeling ways brought him to the attention of the virulent FDR-hating publisher of the *Tribune*, Robert Rutherford "Colonel" McCormick. The Colonel had several axes to grind with Washington, and publishing Johnson's florid dispatch was just part of his maverick campaign against it. He dispatched Johnson to cover the war in the Pacific, and Johnson wound up embarked in USS *Lexington* "CV-16/Lady Lex!" for the action.

The Ship's PAO may have failed to have him sign a secrecy agreement. In any event, Johnson was either shown or had inadvertent access to classified information, and did not view himself as being bound to protect it. It is funny that journalistic ethics have not changed a great deal when there is a scoop available for the taking.

Shudders ran through the Navy Department at the article's publication, and the chilling prospect that the Japanese would recognize the success at penetrating the JN-25 code would be apparent, based on the precise information about the Japanese order of battle contained in the sensational—and otherwise incorrect—article.

Secretary of the Navy Frank Knox leaned on the Colonel to shut down the publicity on sources and methods, and the Colonel reluctantly agreed to spike the story. It is possible that the disclosure, picked up by a couple other major dailies, might have passed without issue if no one made a fuss over it.

The Redman brothers seized on the substance of Johnson's article, which they correctly deduced came from the classified 31 May "Fleet Intelligence Bulletin to Commanding Officers," identifying the disposition and identity of the Japanese forces defeated at Midway.

Eddie Layton, Mac's Boss, goes on in his memoirs to describe the following leaks to legendary radio newshawk Walter Winchell, who made two broadcasts decrying the compromise while explicitly talking about it. The Redmans pushed for an indictment in Federal Court against McCormick and Johnson, managing to keep the matter going, and a matter of public record.

The story broke out again on the 8th of August, 1942. Years later, Jasper Holmes wrote about the impact of the headlines and the following publicity engineered by the Redmans in his book "Double-Edged Secrets." It was published before the ULTRA secrets were declassified, so one has to read the book with an eye to what was *not* said. But as Jasper felt, it was true that "Any informed reader could only conclude that Japanese codes has been broken."

Eddie Layton's 1985 book "And I Was There," lays out a case of staggering mendacity that followed triumph. The Redmans wrote mutually re-enforcing memos up the chain accusing Joe Rochefort of insubordination, and recommending HYPO be brought to heel, and be placed under an officer more to their liking.

The younger Redman, John, managed to get himself assigned to the CINCPOA staff as communications officer and used a private coded circuit belonging to Admiral Nimitz to keep Washington apprised of his progress on isolating the renegade code-breakers.

With all the news of compromised codes flying about, it should not have come of much surprise that the Japanese changed their version of the JN-25 code a week after the news of the *Tribune* indictments, and the work of the previous six months was rendered useless. It would take four months of round-the-clock work to recover the ground that was lost.

Fleet Admiral Bill Halsey always said it was the campaign in the Solomons that was the turning point of the war, not the battle of Midway. I suspect he felt that way because he was not there, being confined to his hospital bed during the fight.

But his point it taken. The see-saw battle to keep Henderson Field on Guadalcanal in American hands gave birth to the ironic unofficial motto of the Marines that the "Navy will always abandon you in a pinch." The Tokyo Express roared in each evening by sea to re-supply the Japanese forces, and before it was done, two dozen men-of-war littered the floor of Iron Bottom Sound. When the battle was over, in February of 1943, the Imperial Fleet never advanced again.

I will ask Mac his professional opinion on whether the single-minded campaign by the Redmans to wage war on Joe Rochefort might have disclosed the success of Station HYPO against the codes to the watchful Japanese.

Joe Redman put on the rank of Rear Admiral, and John made Captain. I understand ambition, but this might be something else. If what they did had caused the Japanese to re-think their security, they might be guilty of something more than careerist aspirations.

You see, the Marines landed on Guadalcanal on the 7th of August, and when the JN-25 codebook changed the next week, the Americans were suddenly flying blind with forces in contact with the enemy. How many people died as a result?

THE WAR IN THE NAVY

• • •

ADMIRAL RICHMOND KELLY TURNER, 1945.
OFFICIAL NAVY PHOTO.

C DR Joe Rochefort had a sign behind his desk at Station Hypo: "There's no limit to what you can accomplish, so long as you don't care who gets the credit!"

That was not true for others in the Pacific. Not the Redman Brothers, certainly, but at least their mendacity is understandable. They were not running the show when the deal went down. Sorry to backtrack on you, but this is necessary.

If you want the man most responsible for the successful Japanese attack, you should not throw a pebble on the grave of poor Husband

Kimmel, who watched his fleet being destroyed in the harbor on December 7th.

THEN-CAPTAIN EDWARD LAYTON. OFFICIAL U.S. NAVY PHOTO.

Eddie Layton was there. He said, "Kimmel stood by the windows of his office at the submarine base… a spent .50 caliber machine gun bullet crashed through the glass." It cut the front of his white blouse and bruised him on the chest. Layton reported the Pacific Fleet Commander said: "It would have been merciful had it killed me."

Kimmel was prescient about that. The cover-your-ass drill began almost immediately back in Washington. Kimmel was sent packing ten days later and Chester Nimitz was brought in. Kimmel would spend the rest of his life defending his actions prior to the attack, accurately pointing out that crucial information had been withheld from him in the crucial months before the disaster.

The real culprits in the failure never paid a dime for what they did, and the culpability went right to the top.

The officer who was directly responsible for the failure of the Navy to be ready was a son-of-bitch named Richmond Kelly Turner. I will say it without emotion at this distance, but in the day, he was the Navy's equivalent of George Patton: serenely confident of his own abilities and filled with a divine certainty of the correctness of his judgment.

He was a tall and imposing man with beetling brows, sharp intelligence and belligerent manner.

He was commissioned a regular deck officer, ranking fifth in the Annapolis Class of '08, and a force of nature. He rose through the Battleship Navy as a hard man, impatient of his subordinates but invaluable to weaker officers who were senior to him. That includes Admiral "Betty" Stark, the Chief of Naval Operations who became the kind of Flag officer that Eddie Layton was fond of saying "couldn't go ashore without giving detailed instructions to the coxswain."

He was possessed of a self-generated vision. He observed that the future of naval warfare involved the airplane, and as a Commander, volunteered for flight training at Pensacola. He later commanded a seaplane tender and served as XO of the USS *Saratoga* (CV-3), one of the first modern big-deck (for the time) aircraft carriers.

Then and now, only rated aviation officers can command what were clearly becoming the queens of the Fleet, so as I said, Richmond Kelly Turner was not a stupid man.

He attended the Naval War College at Newport in 1935, and was kept on until 1938 as the head of the Strategy faculty. He never had a lick of intelligence training, but he was absolutely confident of his ability to craft strategy.

His last ship (he would command task groups as a Flag officer) was the heavy cruiser Astoria (CA-34), and therein lies a tale.

USS Astoria (CA-34) underway off Hawaii, 1942. Official U.S. Navy photo.

Upon completion of exercise Fleet Problem XX in early1939, Captain Turner and his sleek warship were summoned north to embark the ashes of Japanese Ambassador Hirosi Saito for the journey back to his homeland. It was a gesture calculated to express America's gratitude Japan in a period of rising tensions, wrapped in the guise of reciprocity for the ceremonial return of the remains of United States Ambassador to Japan, Edgar A. Bancroft, who died on post in Tokyo, in 1926.

Brief stops for fuel and ceremony with local Japanese communities were conducted in Panama and Honolulu before proceeding west across the wide Pacific. On 17 April, escorted by IJN destroyers *Hibiki*, *Sagiri* and *Akatsuki*, Turner steamed slowly into Yokohama harbor with the United States ensign at half-staff and the Japanese flag at the fore.

A 21-gun salute from Astoria was returned by the light cruiser *Kiso*. American sailors carried the ceremonial urn ashore that afternoon, with a state funeral held the next day. After the ceremony, the Japanese turned on the hospitality for Turner and his sailors.

I have seen a picture of one of the parties that were held in honor of the visiting representatives of the Main Enemy of Japan. At the Tokyo Naval club party on April 19th, Captain turner is seated in the front row, just a few seats away from then-VADM Isoroku Yamamoto, IJN, the architect of the strike on Pearl.

ADM Stark as CNO. Official U.S. Navy photo.

Eddie Layton used to play bridge with Admiral Yamamoto, when he was a language student there.

Captain Turner radiated charm and was praised by U.S. Ambassador Joseph C. Grew for his grace in the diplomatic process.

The whole visit went so well that Ambassador Saito's widow donated a pagoda to adorn the yard of Luce Hall at Annapolis.

Astoria departed Yokohama for a round of good-will port visits at Shanghai, Hong Kong, Manila, and Guam before returning to her home port at San Pedro. Clearly earmarked for flag rank, Turner reported to Main Navy to become Director of War Plans (OP-16), working for the 8th Chief of Naval Operations, Harold Raynsford "Betty" Stark.

"Betty" Stark got his nickname as a plebe in the Class of '03, and you have to put him down as the other major enabler of the disaster at Pearl. An intelligent and insightful officer with a tousle of gray hair, Stark hated controversy and was grateful that the forceful Turner was able to take over the tough and mind-numbing job of generating the detailed plans that would be used to take the war to Europe and Japan.

The problem was that his portfolio in Op-16 had two parts: *plans* and *estimates*. The former would determine how the coming war would be waged. The latter contained the critical elements of where and when. I told you Kelly Turner was a son-of-a-bitch earlier, and what is more, once he was out of his area of expertise as a line officer, he was wrong more often than he was right but incapable of admitting it.

Accordingly, when his Office didn't like the assessment from the Office of Naval Intelligence, he directed it to be changed. He used his Flag rank and access to Betty Stark to bulldoze all opposition. He seized control of the Naval Communications and the products of the Fleet Radio Units and wrenched the analysis over to his estimates section.

There, he had three officers preparing the assessment of what the Japanese were going to do. They were not intelligence officers, but they did know what their Boss wanted.

The war in Main Navy was as savage as anything that happened in the jungles of the Pacific later, and the graves of thousands of sailors and Marines from Pearl Harbor on are directly attributable to the staff wars that went on in Turner's time at War Plans.

Those are bold statements, I know, and the heavy secrecy that wrapped the ULTRA program enabled those who won the staff war and lost a Fleet on December 7th were able to pin their mistakes on others.

Thankfully, we say, it can't happen again. We learned our lesson, right? Remember the sign over Joe Rochefort's desk. I do.

When I was ending my career in the Navy, I watched the new Administration of George W. Bush come to the Pentagon. Uncle Don Rumsfeld was a bureaucratic bully like Terrible Turner. He brooked no opposition to what he already knew to be true.

When the intelligence gang would not go along with revealed truth, he entrusted his Plans and Policy Chief Doug Feith to set up his own little analytic office to cull through the raw material to find nuggets that supported the position Uncle Don was convinced was right.

You know where *that* went.

Oh, and one other thing. In a stunning reorganization, the Navy has consolidated all of its "information" resources into a

New OpCode. The former Radio Intelligence tribe that got de facto independence after World War Two is being jammed together with the Office of Naval Intelligence and Naval Communications. The CNO is making an intelligence officer walk point on the reorganization, but there is clearly going to be a sharing of leadership in the future. We will clearly someday have a Communications officer running the show. It is not 1940 all over again, I am sure.

But it certainly is back to the future, isn't it?

GRAVE OF ADMIRAL R.K. TURNER AT SAN BRUNO.

The secrets that Eddie Layton and my pal Mac kept finally came out in the 1970s, and Mac led the drive to get Joe Rochefort a posthumous Distinguished Service Medal. Eddie Layton died before completing his book, but it was finally published in 1985, giving the first account of how badly the Navy leadership had botched the analysis of Japanese intentions.

Kelly Turner died with his reputation intact on February 12, 1961 He is buried in Golden Gate National Cemetery in San Bruno, CA, alongside his wife Hattie, and near those of Fleet Admiral Chester Nimitz, Raymond A. Spruance, and Charles A. Lockwood with their spouses.

It was an arrangement made by all of them when they were alive.

Spruance had his nomination to Fleet Admiral blocked by Congressman Carl Vinson, who got a Nimitz-class carrier named in his honor for serving as the Chairman of the House Armed Services Committee. Vinson preferred that Bill Halsey get the honor. Lockwood was the architect of unrestricted submarine warfare that strangled the Home Islands.

And of course, after Kelly Turner was eased out of War Plans due his belligerent inability to work with the Army, he led the naval campaign at Guadalcanal. That was the one that began just as the Japanese changed their codes, and Kelly and his sailors and Marines had to fight in the blind.

What goes around, you know? But how many of them had to die for Turner's obstinacy at War Plans?

BRANCHES AND SEQUELS

• • •

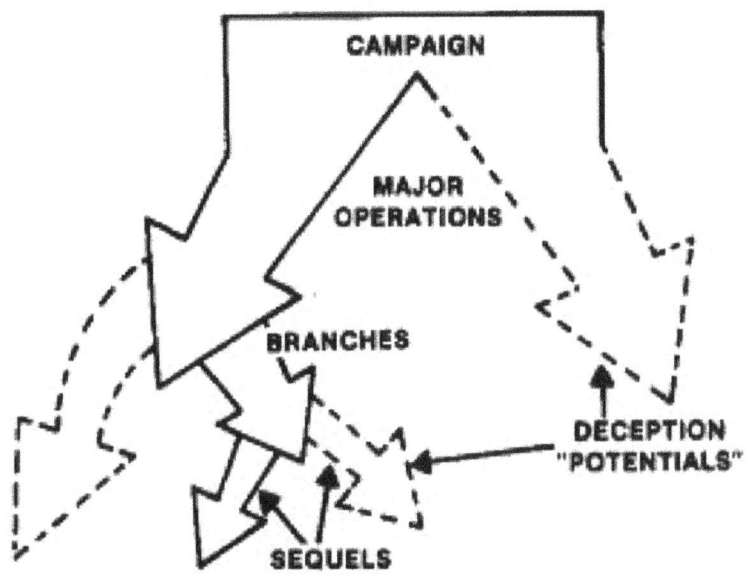

MILITARY DOCTRINE OF "SHIT HAPPENS" FROM JOINT PUB 1.

Preparations for offensive operations against Japan had been in place since the pre-conflict days of Plan Orange; there is nothing our military has learned how to do better than plan.

Mac was part of all that, of course, and I have a couple of stories that Jasper Holmes used to tell about him.

The mission was spilling out of the Dungeon at Building One at the Pearl Harbor Shipyard and was changing as the requirements to support offensive amphibious operations in the Southwest Pacific emerged. It was in The Plan.

Six weeks after the Battle of Midway, the Intelligence Center Pacific (ICPOA) was established with CDR Joe Rochefort temporarily double-hatted as Officer in charge of both ICPOA and the operationally subordinate Station HYPO.

The first U.S. doctrine of expeditionary warfare came with the development of War Plan Orange in 1890, long before the little war that stripped Spain of its possessions in the Caribbean and Pacific, and catapulted America into the ranks of the Colonial powers in 1898.

ANNEXATION CEREMONY AT THE IOLANI PALACE, HONOLULU, 1898.

Theodore Roosevelt modified War Plan Orange as a contingency for conflict with Japan. The annexation of Hawaii in 1898 relieved the immediate tension with the burgeoning Meiji empire, but the Navy and War Departments continued to plan.

Generally speaking, the Plan was inviolate, and was based on enduring principles and assumptions from the Spanish War that were a half-century old.

As rising senior officers in the joint schoolhouse, my generation was taught to deal with reality the Army way, with the going in assumption that "no plan survives first contact with the enemy," which is military-speak for the civilian doctrine that "shit happens."

To deal with an altered landscape after that first contact, alternative scenarios were always considered. We were taught to call these "Branches and Sequels" as consequences flowed downstream from events.

There were other plans, of course, a whole rainbow of colors. PLAN RED was the Atlantic Strategic War Plan, which originally was oriented against Great Britain, if you can imagine, though branches and sequels in 1914 resulted in the substitution of Germany for the UK. The First War branches and sequels created the RED-ORANGE PLAN, which hypothesized a two-theater war with initial emphasis on operations in the Atlantic.

War Plan Orange, which came in Navy and Army flavors, consisted of three phases, adapted as new technology changed military capability being introduced to the Army and Navy.

Plan Orange envisioned:

Phase I: The U.S. expected the loss of the lightly defended outposts south and west of Japan, which could not be defended. The Plan assumed the Navy would concentrate the Fleet at their homeports in order to surge forward.

Phase II: With superior naval and air power, the Navy and Marines would advance west. Small-scale attacks against Japanese-occupied islands would capture them and establish supply routes and overseas basing for new long-range bombers like the B-29.

Phase III: The U.S. would then advance toward Japan utilizing islands that were parallel to and near Asia. These newly acquired bases could choke Japanese trade and allow air bombardment of Japanese cities and industry, leading to victory without invasion of the Japanese homeland.

The complex nature of the plan required close cooperation between the Army and Navy, and Army-Navy Board (better known as the Joint Board) was created in 1903 to de-conflict the efforts of two independent cabinet Departments.

We talked about Richmond Kelly Turner, whose high-handed and arrogant conduct as Chief of Navy War Plans (OP-16) was not limited to abusing the Office of Naval Intelligence. His contempt for the Army was so profound that George Marshall demanded he be removed from his position.

CNO "Betty" Stark had no stomach for a fight with Marshall, and despite Turner's success in devising a Rainbow palette of options based on Stark's "Plan Dog" memo of 1940.

While he might have been indecisive, Stark was intelligent and subtle. He anticipated an expected two front war against Germany and Italy in Europe and Japan in the Pacific, with Europe the first priority.

Turner was dispatched to cool his heels as a Deputy Chief of Staff under leathery Admiral Ernie King as the newly-created Commander in Chief U.S. Fleet for six months.

Phase II of the plan provided a combat job for Turner as Commander of Amphibious Forces Southwest pacific that could harness his truculent intelligence into something useful.

The Branches and Sequels created by the code-breakers who enabled the victory at Midway let Phase II of Plan Orange go forward. The plans existed for operations in the Solomons, and the Marines landed on Guadalcanal.

Phase II required Eddie and Joe and Jasper and Mac to adapt to new Branches and Sequels, and that meant that the nature of intelligence support to the war had to change.

Part of going on the offensive was using the radio intelligence to interdict and foil Japanese operations while not disclosing the sensitive nature our of sources and methods. Which is how Eddie Layton came up with a novel plan to kill his old bridge partner, Admiral Isoroku Yamamoto. That is quite a story in its own right, but as they say, fortune favors the bold, right?

GAG RULE

• • •

Help yourself to a piping hot cup of
STOP ASKING QUESTIONS
and you'll worry a whole lot less

After all, you're a Citizen... right?

There are a whole bunch of good reasons why the memo came down saying that I could no comment on it. You know what I am talking about, though I am not talking.

I thought it was a fairly reasonable bit of coverage, probably something the public ought to be aware of, but according to the memo, I can't even say what it is.

I support the memo, and I support the right of the Press to talk about whatever they want in this enduring constitutional democracy. I remember the last time I got a memo like that. A book about the submarine force was issued, and some retirees had let their tongues wag fiercely about the things they did and the places they did it.

The Brass was alarmed and told us that secrets remained secrets. Even though the stories were out there in black and white, we were supposed to not talk about anything in the book. I bought it, of course, just to see what the authors had right and what they had wrong. I was impressed by the level of detail they had accumulated and dumped out on the prim white pages.

What the account lacked in strategic context it attempted to make up in sensation, just like what I am not supposed to talk about did, not that I am commenting on it.

Heaven forbid. Mac would understand perfectly well. We live in an elliptical world. Much of what we talk about at Willow, and will talk about again next week, was once fiercely classified.

The Great War against the Fascist powers forced all sorts of innovation. If you want to understand our world today, it is simple enough to go back to the secrets of yesterday and take my assurance that not a great deal has changed.

Sources and methods evolve, of course, but the way we work and the way we attempt to keep things secret do not. The article I am not talking about at the moment is clear enough about that, and even if the topic seems to be startling and sinister, the secrets are not nearly as big as other things.

I got a note, for example, from a colleague who wanted to know about the Redman Brothers. Who were those guys, he asked, and what did they do? I sighed when I wrote back. Joe Redman was a Rear Admiral and Director of Naval Communications twice, I wrote, and his little brother John was a Navy Captain. They both made their careers on the great victory at Midway in 1942, and they stole the credit from Joe Rochefort.

Then they had the *real* hero relieved in the manner of an NYPD Detective in Manhattan who is put back in uniform and sent to walk a beat on Staten Island. Imagine it; the best Japanese linguist and code-breaker in the Navy dismissed from his post, and placed on a drydock for the duration!

The perfidy of the Brothers was concealed by the 25-year gag rule on ultra-top secret of the ULTRA program. It was not until 1970 that the archives were cracked open on the now-ancient war, and while the historians were agog, the rest of the world had moved on.

Here is how the Redman brothers did it.

In separate memoranda to the Director of Naval Communications on 20 June 1942, less than three weeks after the victory at Midway, each of the Redmans criticized the work of Joe Rochefort and Eddie Layton. "Remember," said Mac, "these were the guys who said he Japanese attack would come against Dutch Harbor in Alaska. If Admrial Nimitz had believed them, we would have lost Midway Island, and the Japanese would have consolidated an island perimeter that would have been hard to crack."

The senior Redman's memo snuck up on the real issue. After several paragraphs justifying why Radio Intelligence (the unclassified euphemism for ULTRA) should remain under *Communications* control rather than in the Office of Naval Intelligence.

Joe Redman wrote this about Rochefort and Layton: ". . . they just don't speak our language. The intercept material must be obtained by operators trained in the Kana code. The source of the operators is Naval Communications . . . the intercept equipment belongs to Communications . . . the question of traffic analysis involves personnel and only those familiar with radio communications can properly administer this work."

Captain Redman then got down to the real business of his memo, which was the personal destruction of the men who cracked the Japanese battle plan. "(Rochefort and Layton) are not technically trained in naval communications, and my feeling is that radio traffic analysis, deception, and tracking, etc., are suffering because the importance and possibilities of the phases of radio intelligence are not fully realized.

I believe that a senior officer trained in radio intelligence should head up (a Radio Intelligence unit) rather than one whose background is Japanese language."

To put the finishing touches on the matter, Joe Redman's baby brother John signed a letter that pounced on a formal request from Admiral Nimitz letter of 28 May 42 addressed the "inadequacy of the present intelligence section of (my) staff."

Admiral Nimitz wanted additional resources to be placed under the intelligence department he already had, and he fully supported Eddie Layton, Jasper Holmes and Joe Rochefort. But he gave Washington the chance to twist his words. Rochefort did not get the medal he earned by handing Nimitz the greatest victory of the Pacific War.

No one could talk about what happened for twenty-five years, due to the gag order on the Big Secret. Mac looked over at me the other night and said that Captain Goggins showed up to replace Joe Rochefort.

"He was one of the Redman Brothers home boys, wasn't he?" I asked. "Yep."

"How did you work for him under those circumstances?" I said. "That was outrageous, and the office politics in Washington could have cost American lives!"

"We had to go on. There was a war to win," said Mac, a little wistfully. "Layton and Holmes survived the coup, and Joe was the sacrificial lamb to the Redman Brothers ambition. By the time Eddie Layton could talk about what happened, the official story was already written. There is a building named for Joe Redman over at the Nebraska Avenue complex. I used to see it when I worked there after the war, and all I could do was mutter under my breath."

That is all I can do about the other thing that I am not talking about, but I am a good sailor, and I can follow orders with the best of them.

The hubbub at the bar had a sort of desperate merriment, like people were trying to forget the economic news, and the new guidance from Secretary Gates to trim Defense contracting. It made it hard to hear above the wine-fueled din.

I was determined to get through the big transition in the intelligence organizations that happened on Hawaii after the victory at Midway in 1942 and I was making slow going of it. It is not as exciting as the magic moment at Midway, but that was built on months of mind-numbing analysis.

Plus, the food at Willow was too good, and there is a lot of other stuff going on to talk about.

"OK, I said. "It is 1942, the war is being taken—barely—to the enemy, as that stubborn jerk Richmond Kelly Turner landed the Marines on Guadalcanal, and then cut and ran on them."

RADM RUSSEL WILLSON RELIEVES CHESTER NIMITZ AS BATDIV 1 ON USS *ARIZONA*, 1939. OFFICIAL U.S. NAVY PHOTO.

Mac nodded, patient with my disorganization.

"Turner had a lot of baggage on the war being the way it was then," I ventured. "His stubborn determination to do his own analysis was a lot like Secretary Rumsfeld," I said taking a sip of white wine at the little table, moving the Senate notepad around a stack of books and the plates that went along with the fish-and-chips.

"His control of Radio Intelligence and the decision to withhold information from Admiral Kimmel and his intelligence Officer Eddie Layton at Pearl unquestionably contributed significantly to the disaster," replied Mac.

"Then, the revelations of the Redman Brothers, who were determined to destroy Joe Rochefort . . ."

"And Eddie," Mac reminded me. "Admiral King wanted Nimitz to fire him, too, but he wouldn't. Another villain in all this was Admiral Russell Willson—two ll's—who was Chief of Staff to Ernie King back in Washington. He had a real mean streak and listened to the lying Redman brothers."

"Could it have something to do with the fact that Admiral Nimitz was advanced ahead of him to command the Pacific theater? I understood Willson relieved Nimitz on *Arizona* before the war."

"Could be. But he was a son-of-a-bitch. The Redmans used him to pursue the leak of the code-breaking revealed in the Chicago papers until it was all in the papers again. They had to have a Grand Jury and a publicity circus."

"The Japanese may have got the hint that way, or it could have been something else. But if the codes were changed because of those bastards, four American cruisers were sunk in The Slot because there was no warning available for them." Mac furrowed his brow, and his eyes misted at the loss of hundreds of sailors on those proud ships.

"The change in the JN-25 code system meant that Station Hypo had to start over, almost from scratch. The Marines in the jungle and the fleet at sea were in the dark except for what they could see themselves."

I scribbled madly. "OK—now, you got to that point still working in the basement?"

"Yes," said Mac. "We were growing. We had a space over in the Supply building. That was mostly the Air Intelligence folks coming in, but things were definitely growing. In response to a Marine Corps requirement, Admiral Nimitz directed the establishment of the Intelligence Center Pacific, or ICPOA, about six weeks after the battle."

"Joe Rochefort was the first Officer in Charge, right?"

"Yes. That was before the summons came from Washington for a short Temporary Duty period. We all knew he would not be back. He turned over the keys to his desk before he left. At the time, he was dual-hatted as the OIC of FRUPAC and ICPOA. "

"Jasper Holmes wrote that there really wasn't any such thing in the early days." "He was quite right. In the beginning, Station HYPO took the cover name "Combat

Intelligence Unit" (CIU) to deflect speculation about the sensitive work performed there.

As more people arrived to augment HYPO, those not engaged in strictly Radio Intelligence functions were assigned all-source analysis duties under Jasper, involving enemy ord. That group assumed the CIU name, while still located in the same place. "It was a cover, then."

Mac chuckled. "Yep. That meant there were some other issues. I was a deck officer, remember, and the Bureau tried to send me to sea twice since the Fleet was growing like Topsy. They wanted me to go on a minesweeper, but it was a case of me knowing too much about something the Bureau didn't. Jasper managed to get me ordered to the Hawaiian Sea Frontier, but I never changed my desk."

"That was about the time the official name FRUPAC actually was used, right?"

Mac nodded. "Later on, after Joe left and Captain Goggins showed up to relieve him. He got there before the orders did, and we thought he was supposed to command ICPAO, not replace Joe."

Mac looked off in the general direction of Peter, who was doing one of those graceful pours from a bottle into a tulip glass at the amusement of two attractive ladies who were seated close together at the bar.

"Goggins was no cryptologist," He said. "he had some communications background, but he was really a line officer, which made the lies of the Redman brothers that slandered Joe so hard to take. Goggins had been XO on *Marblehead* at the Battle of Makkassar Strait, and he was badly burned when the Japs hit her. He was still recovering when he reported."

I looked at my notes. "They added a "J" to ICPOA to reflect Admiral Nimitz's role as the joint force commander," I read.

"Correct. Eddie Layton found an Army topographic unit over at Fort Shafter that wasn't doing anything, and to accommodate the new composition of the unit, they ordered in Colonel Twiddy. He was the outside guy, always with a big cigar. Jasper Holmes would work at the Estimates section in the morning, and then spend the afternoon as the XO at JICPAO. Jasper did all the heavy lifting."

"And what happened to CIU?"

"It became the Estimates Section, organizationally part of the JIC, but remaining in the same building with Radio Intelligence, using their decrypts and traffic analysis to formulate an all-source intelligence picture for the daily onion-skin overlay for Admiral Nimitz plot, and the weekly Intelligence Bulletin."

"Did they report to Eddie as the Fleet Intelligence Officer?" I asked.

"Nope. FRUPAC always reported to the CINCPACFLT Communications Officer," said Mac.

"Isn't that a bitch!" I exclaimed. That was that asshole, John Redman!" How did Eddie put up with him? You must have known him."

"Nope. I was not on the CINCPACFLT staff until we went forward to Guam. John Redman stayed in Hawaii."

"That doesn't surprise me in the slightest," I said, putting down my pen and raising my empty glass to signal Jim behind the bar for reinforcements.

"I don't know what Eddie thought about him," said Mac. "But he doesn't appear anywhere in his book about the rest of the war, except to note that he was probably responsible for the change in the codes.

"And the deaths of a bunch of sailors in The Slot," I said. "No wonder there has been bad blood between the Cryppies and the Intel guys for all these years. What a story."

"We ignored that, and most of us were not intelligence anyway, at the time. But let me tell you how it worked when Eddie shot his poker buddy Yamamoto."

I could see there was more wine on the way. It was a good thing because this was going to be a *hell* of a story.

SEATED UNDER A TREE

• • •

ADMIRAL ISOROLU YAMAMOTO, 1942
PHOTO: WIKICOMMONS

B ig Jim filled my wineglass to the precise point in the Tulip where the air could mingle sufficiently with the ambient air to produce a delightful aroma while not encouraging the alcohol to evaporate. I was getting to the point that I did not care, pleasantly warm in all my appendages, but the ritual was a comfort.

Mac contemplated another Virgin Mary, which could only evaporate the essence of tomato and olive, and decided he had enough vegetables for the day. I knew we were about to draw to the end of this session at Willow, so I underlined the notes about Pearl getting surprised again—or better, warnings ignored—and asked the question.

"Well, they lost their Empire," said Mac. "I suppose we have to get used to it as well, and hope to do it with the same sort of grace."

"I don't know," I said. "They want to call that new building in New York the Cordoba Center. Putting up with a Mosque at Ground Zero named for the capital of Islamic-occupied Spain seems pretty darned tolerant, don't you think?"

Mac just smiled, and shook his head.

"They found the Admiral seated under a tree, from what I have read."

Mac nodded, shifting gears from his earliest months in Hawaii to nearly a year later. A fresh new-guy fish in the Dungeon basement under Joe Rochefort and Jasper Holmes to a seasoned analyst, thousands of dittoed cross-indexed files, looking at other fresh fish arriving each day with veteran eyes.

"Yes," he said. "There are conflicting reports about it. Some say the initial Army patrol to the crash site found Isoroku Yamamoto still in his seat, very peaceful-looking, holding his *kitana* sword by the hilt. Thrown clear of the wreckage of the Betty bomber, they say.

Other accounts had him with a couple flesh wounds. But he was dead, all right. Maybe shock."

"They did not try to recover the body?" I asked with surprise.

"They sent a Navy patrol the next morning, at first light. There was a great deal of service deference. The Army connection was useful," Mac said, swirling the ice in the bottom of his highball glass. "The message notifying the local commands of Yamamoto's inspection visit was passed to the Army in a less secure code than the Naval JN-25, and the Japanese assumed that was the way we were able to intercept the flight. The katana and the insignia of rank were missing by the time the Navy patrol got there. Someone got some great souvenirs, but they have never turned up."

"I will look on eBay," I said. "This was April of 1943, right?"

"The 18th," said Mac. "But we had been watching the traffic out of Rabaul very closely for months. We recovered all the traffic about the Combined Fleet moving its headquarters, and the comings and goings of fleet units in and out of Rabaul. The Harbormaster was a big help to us. He was meticulous about reporting everything that moved, and the merchant ships were on a lower-level cipher system that enabled us to recover the unit identities for some of the combatants.

"Keep Cool, fool, it's Rabaul." I recited the old rhyme like a modern rapper. The Admiral smiled. "It certainly was, then."

"So, I understand there was a relationship between Eddie Layton and Yamamoto. His book recounts a bridge game."

"There certainly was, but he told me it was poker, not bridge. The Navy established a Japanese language program way back in 1910. Under the provisions of the plan, two officers a year who had completed five years of sea duty were sent on independent duty to learn the language. Eddie was sent in 1929, after serving his time on the destroyer USS *Stack* (DD-406)."

"Joe Rochefort was there at the same time, right?"

The Admiral nodded. "That is where they met for the first time, and that was one of the lies that the Redmans spread, that Joe was not a qualified linguist. He most certainly was."

"Eddie actually knew Yamamoto, didn't he? That makes the whole thing sort of creepy."

"We didn't look at it that way. "Terrible" Turner had his picture taken with Yamamoto when his ship called there. Eddie actually knew him pretty well. Yamamoto had participated in the Japanese equivalent of our language program since we both knew there would ultimately be a show-down in the Pacific, and some familiarity would be useful in killing each other. Eddie first came in touch with the Admiral when he was Naval Vice-minister, and the search for Amelia Earhart was going on."

"Did the Japanese kill her and her co-pilot Fred Noonan?" I asked.

"Won't ever be known, for sure. Certainly they were very defensive about what they were doing in the former German colonies, pouring all that concrete."

"Eddie and Isoroku were on a first-name basis. He told me they attended an evening of kabuki, which Eddie loved. Yamamoto seemed to appreciate Eddie's interest in the cultural life of Japan, and mostly they stayed away from business."

"Wasn't there a strict prohibition on the Americans doing anything like intelligence collection?"

"Absolutely. The Japanese militarists were clamping down hard on security, and placed whole distracts off limits to foreigners. But Yamamoto was always correct. He hosted a duck-netting party in the late thirties—maybe 1938—for the foreign attaches.

It was one of those Japanese things where the outcome is ordained, and there will be duck sukiyaki whether everyone had netted a duck or not."

"I have been to those," I said, referring to my notes. "Eddie's book says it was three rubbers of bridge after dinner with the Admiral, and he won all three. They were drinking 'John Begg' whiskey in square ceramic jugs."

"It was Johnny Walker Black in my days in Yokosuka that the Japanese liked."

Mac shook his head. "I always heard from Eddie that it was poker. He loved the game, and was a skilled player and usually won. I should tell you about his mission in China to play cards with the aircrew of the Warlord Chang Hsueh-liang's Ford Tri-motor to get intelligence on his travel itinerary."

"I would like to hear about that, but for now, the story about Eddie and the Admiral seems like it has a real personal element."

"I suppose it did. I can only imagine what he really thought when Jasper called him on the secure line from the Dungeon and told him what we had. Jasper brought me the original message and told me to plot it out to see if it made sense, and our decrypts of the place names were correct."

"NTF131755 was the message," I said, peering at my scribbles, "and it was addressed to the commanders of Base Unit No. 1, the 11th Air Flotilla, and the 26th Air Flotilla."

"Yes, it was copied by our Hypo personnel and two other stations in the Pacific." Mac remembered it almost verbatim, and recited it, almost in a trance:

"On 18 April CINC Combined Fleet will visit RYZ, something something. In accordance with the following schedule:

"Depart RR at 0600 in a medium attack plane escorted by 6 fighters. Arrive RYZ at 0800. Proceed by minesweeper to somewhere else arriving at 0840."

At each of the above places, commander in chief will make a tour of inspection and at unknown location he will visit the sick and wounded but current operations should continue."

"That sounds like there were a lot of holes in it."

"Yes, but Dick Emery managed to identify some of the outlying fields, and things began to come together. Jasper knew it was significant, and after he talked to Eddie, he was told to bring it with the plot I did to the Headquarters at the Sub Base."

"Didn't Justice John Paul Stevens work on the message?" I asked. "That is a remarkable part of his biography."

"There were a lot of lawyers in the intelligence billets," said Mac. "Stevens had just the sort of skills that Forest Sherman was looking for when he set up the air intelligence program.

As far as Stevens being part of the decryption team, he was in Estimates with me, which is when we used to go to lunch at the Makalapa BOQ. But I don't remember him being there at the time. I could be wrong. There are a lot of people who took credit for what happened next."

"You don't seem to be wrong about much," I said.

"There are some things that just need time to get straight," said Mac firmly. "And what happened to Admiral Yamamoto just gets to the point. It is like the Admiral told Eddie after the poker game: "Science and skill will always win over luck and superstition."

Turns out the Admiral used the old Chinese trick of drinking water instead of whiskey."

I sighed. Maybe I should try that for these sessions with Mac. I shrugged. Too late now. This was going to take another glass of wine, and my notes were starting to soak up the moisture from the bottom of the tulip glass. I waved to Jim at the bar and signaled for more.

FIRST TO THE BLACKBOARD

• • •

Its better to have an
Army of deer being led by a lion,
rather an Army of Lions being led by a deer . . .

What we do is what we do.

Jasper Holmes called Eddie on the secure line and told him what they had about Admiral Yamamoto's schedule. Mac was told to plot it out, make sure the recovery fo the place names made sense. It seemed to, and the decrypted message and the chart went over to the headquarters.

Then everyone went on to what they did, and what we all have done for all these years. Listening. Copying. Thinking.

Chester Nimitz thought about the impact of killing the architect of the Pearl Harbor attack, and the architect of the assault on Midway that brought disaster to the Combined Fleet and the loss of four fleet aircraft carriers, all their aircraft, and above all their precious cadre of pilots. Pluses and minuses. We do what we do.

Nimitz coordinated with SecNav Knox, who asked the President.

I don't know for sure, but it is possible that FDR inserted a Chesterfield into his ivory holder and lit up before he nodded. With that, the mission to kill Fleet Admiral Isoruku Yamamoto began to roll.

The leaders weighed it against the risk of compromising the penetration of the JN-25 code system and then made the decision.

Kill the bastard.

Eddie knew the man, had played cards with him for modest stakes. In some ways, they were the mirror images of one another. Yamamoto the man, sent to America to soak in the language and the culture, Eddie, sent in the other direction to learn the all too scrutable East.

To kill the man and keep the secrets, they first needed some plausible deniability. The fiction that an Australian Coastwatcher had passed the information on the whereabouts of the Admiral of the Combined Fleet was passed to rambunctious Admiral Bill Halsey in the Southwest Pacific. The precise timing of Yamamoto's flight, escorted by nimble Zeros and accompanied by another Betty bomber with his chief of staff embarked, was too precious for words, an event in space and time in the future covered by a fiction from an imaginary past.

We have all been there in our line of work. "You had a pretty good handle on the code designators, right?" I asked Mac. "I mean, you were able to plot it out, from the partials in the recovered message?"

Mac nodded. "RR being Rabaul, the fortress island, RXZ was Ballale, and RXP was Buin on the southern tip of Bouganville. The first part of the itinerary looked like it might be in theoretical range of Henderson Field on Guadalcanal. No Navy planes, of course. They did not have the legs for the mission."

The Air Corps had a requirement for long-range escort of the bomber force in Europe, and had produced the remarkable P-38 Lightning, a twin-boomed long-legged two engine fighter that the pilots of the Luftwaffe called "Der Gabelschwanz Teufel," or The Forked-Tail Devil. The Japanese aviators called it "Two planes, one pilot."

My former uncle-in-law Joe flew one out of the Aleutians in the two-thousand mile war there, a Texan in the wild swirling arctic weather of gales and clouds, and he said he was thankful for having two engines and drop tanks and plenty of vengeance in the nose.

The placement of the twin engines on either side of the cockpit meant that the nose of the aircraft could bristle with machine guns and cannons unobstructed by the arc of the propellers: a Hispano 20 mm cannon with 150 rounds, and four Browning 50-calibre machine guns, each with 500 shells in the magazine.

I gestured at the copy of Eddie Layton's book on the table. "This is what you had in the decrypted message: Yamamoto planned to depart Rabaul at 0600 and land at Ballale Airfield at 0800. Then, proceed by subchaser to Shortland at 0840, then depart at 0945 aboard the same sub-chaser and return to Ballale at 1030, then depart at 1100 by G4M1 Betty and arrive at Buin Airfield (Kahili) at 1110. Finally, at 1400, depart Buin Airfield (Kahili) by G4M1 Betty and arrive back at Rabaul at 1540."

Mac nodded. "Yes, but with any luck, he would never land at Ballale."

"Jasper had me plot it out to ensure that we had the places correct, and that the plot made sense. It did."

I wrote that down on the notepad. "Jasper did not take a lot of credit for his role in this," I said. "I read the passage on the shoot-down, and it seems like he barely had anything to do with it."

Mac looked at me kindly. "He was literally the first to the blackboard, and Jasper was a gentleman, unlike some others in this story. The thing to remember about his book is that he was the first to publish anything about the secret history of the. war NSA reviewed his manuscript just after ULTRA was declassified after twenty-five years. It was a bit surprising at the time since they said the whole manuscript was good to go, except they said he could not mention the code designation JN-25. For some reason, the people at the Fort thought it was still sensitive. Of course, by the next year, everyone was writing about the code system specifically, so you have to place the accounts in the context of the year they were published.

"So that is why the official history is wrong," I said. "When I was in a fighter squadron, I learned that the first ones to debrief the mission got to establish what the truth was, and who won and who lost."

"You bet. It is a matter of when people are free to talk about things. When Samuel Elliot Morrison did his multi-volume history none of this information was available. All he said in his account was that Admiral Nimitz and Halsey "learned of the visit.""

Jim passed by and looked at me inquisitively. I glanced at the level of wine in my glass and shook my head. I didn't have far to navigate to get back to Big Pink, but in my old age I am taking fewer risks with the cops.

I signaled to the bar and called out: "Check, when you get a chance, Jim."

"You got it, Vic," he called back, topping off the glasses of the attractive women at the bar.

Mac resumed his story. "A thorough, detailed briefing including the cover story was provided for the pilots, but they were not specifically briefed that their target was Admiral Yamamoto. 16 P-38-G Lightnings were tasked with the long-range intercept mission.

According to my plot of the message, they would have to fly 435 miles from Guadalcanal to catch the Japanese."

"That is a remarkable gamble," I said, fishing for my wallet. "Was it worth it?"

Mac laughed as Jim slipped the black folder with the tab on the edge of the little table, wedging it against the copy of Jasper's Holmes book.

"More than you know, Vic. They had to fly at extremely low level the whole way to avoid detection-less than fifty feet above the waves. The cockpit in the Lightning could not be opened, and you can imagine in the tropics it was hot. The pilots normally flew in shorts and sneakers because of the heat."

"The Lightnings met Yamamoto's two Mitsubishi G4M "Betty" fast bomber transports and six escorting Zeros just as they arrived at Empress Augusta Bay. The Americans split into two groups as the Zeros spotted them, and first Betty with Yamamoto dove to the treetops. The second turned seaward."

"That must have been some pretty intense adrenaline for everyone."

"Don't know. We were on to other things then, back in Hawaii. But there is an interesting story about what happened later."

I slipped my Visa card into the black folder and chose not to look at the tab. Willow takes care of us pretty well, and whatever they thought was fair was OK by me.

"And that was?"

REX BARBER AFTER RECEIVING THE NAVY CROSS.

"The first lighting to recover at Henderson Field was flow by a Major named Thomas Lanphier. He claimed credit for the kill on Yamamoto's Betty. It was bullshit, but he was first to the blackboard on the mission debriefing. Yamamoto was actually killed by Rex

Vic Socotra

Barber, who had sole credit for it. The Admiral was thrown from
the aircraft still in his seat. The autopsy indicated that he might
have survived the crash, since he had no visible wounds aside from
a small cut above his eye."

"Now, none of that was known until long after the war, right?"

"True. We were back to doing what we did. We didn't know for
sure what happened until May, when news of Yamamoto's death was
officially reported to the Japanese press. In the meantime, Barber
and Lanphier were both awarded the Navy Cross, if you can imagine
a couple Air Corps pilots wearing them.

FISH AND CHIPS

• • •

There were a ton of people from the Company at Willow. They clogged the passage past the near end of the bar.

They were not my division, so I didn't know any of them, and they seemed happy enough probably not knowing what Secretary of Defense Gates effectively did to the whole vendor community this week.

I don't imagine they are going to realize until a couple quarters down the road, so I smiled thinly and edged my way through the crowd.

Mac was already seated at one of the little tables across the aisle from the bar. I snagged the seat next to him.

"Someone from your company must have let the cat out of the bag," he said with a merry twinkle in his eyes.

"Face it, Admiral," I said. "You are a rock star!" We laughed, and he handed me a pad of notepaper.

"I thought you might like this," he said. I looked at the letterhead, which read: "United States Senate" in the fancy old English script, sub-headed "Committee on the Armed Services" with tiny letters at the upper left, saying "Strom Thurmond, Chairman."

I gave a low whistle. "I met the Chairman a few times. He would cruise around on his own, even when he was in his late 90s. I introduced him to my parents one time when I had credentials to be on the Hill and played the big shot when family came to town."

"He was an active fellow. I swiped the pad when I was testifying up there on the years ago. You know that he landed at Normandy at D-Day in one of those gliders. He was authentic, regardless of what you thought of his politics."

"I heard in his later years he would go to the buffets and put meatballs and shrimp in his pockets. His staff was appalled. I didn't see anything leaking the last time I saw him, though."

Mac laughed, and I saw the long line of people he had known

and with whom he had interacted. He handed me a pen to go with the pad.

"I thought you could use this instead of napkins," he said.

"Thanks. I am still a little fried from the hours on the road." I blinked from fatigue and the oppressive heat outside. Mac looked cool and crisp as always.

"Let's see, I wanted to talk about 1943, and the last quarter of the year as things changed. We landed on Guadalcanal two months after the battle of Midway with that asshole Richmond Turner in command of the amphibious forces. You don't mind if I call him an asshole, do you?"

"I do myself," said Mac. "The Marines still hate him. We lost access to the JN-25 code right around the time of the landings, and we had no warning to pass when the Japs came down the slot. Turner took off and left the Marines behind, not even waiting to unload the cargo ships with the supplies the Jarheads needed. They don't forget that he cut and ran on them."

"I suppose he had a good argument," I said. "Hard times and hard choices." I waved at Jim, who brought a delightful bottle of Spanish white and a tulip glass that he filled halfway up. I asked him if he could possibly put in a request for the $5 neighborhood bar menu of the miniature fish and chips for us. He said he would think about it and floated off into the crowd of earnest company people.

I gestured at his retreating back with the pen that Mac had given me. "There is a lot of stuff that is going to change around here. I heard on the radio that they are talking about $3.5 billion in defense contract cancellations in Fairfax County alone," I said.

Mac nodded gravely. "That is what happens when things end. He produced a truncated copy of an ancient typed memo. "You asked one time what we did when we got back from Guam after the war ended. A guy writing a book found this in the archives. I don't recommend you go there. All that paper will just suck you in." He gazed at it before sliding it over. "This was in RG-38, Box 94 in the Naval Security Group Archives."

I looked at the ancient paper curiously. It was dated 8 September 1945, and was addressed to the Flag Secretary, and the subject line was "Report of additional Orders and Plans destroyed by Burning." It contained the list of things Mac had made disappear in fire:

1. G-2 Estimate of the 'Enemy Situation in Kyushu 25 April 1945
2. Com3rd Fleet Standard instructions, 1-45 Part one only
3. Command 2nd Carrier Task Force Pacific and TF 38 OpOrd #2-45, 25 June 1945
4. Secret Operations Instructions #88 SW Pacific Area

I whistled. "This is all Operation Olympic stuff, right, the real deal."

Mạc nodded. "That is what the few of us that were left did when we got back. Everyone else went home as fast as they could. We destroyed stuff."

"Were these the only copies?"

"No, but I have no idea where or if it was all kept. That is why people have been arguing about everything ever since. Trust me, you don't want to be in the middle of all those boxes of ancient papers."

"I heard there was only one guy left at the Joint Intelligence Center Pacific Ocean Area," I said. "Twelve hundred people down to a Lieutenant in a couple months."

Mac smiled. "That was LT Wendy Furness," he said. "He was left with two rooms of captured material. Pistols, binoculars, Samurai swords. He was told to get rid of it all and lock up the empty JICPOA building."

"What did he do with it all?"

"Don't know. I got a set of binoculars, though."

"Well, that is a good thing about winning a war. Everyone gets a souvenir, even if it is just your life."

"We are in the process of surrendering in the conflicts we are in now," I said glumly. "And we are so stupid as a nation that we cannot even recognize when someone is building a victory monument in our greatest city that took the biggest hurt."

"What do you mean, Vic?"

"They want to call that new building in New York the Cordoba Center, which was the capital of Islamic-occupied Spain. It is like putting the middle finger up at us and we don't even recognize it." I took a sip of the Spanish wine. "I was listening to an interview with one of the survivors of The Battle of Britain on the BBC this morning. Seventy years on from when he was a junior pilot launching against the Nazis, he is phlegmatic about his role in changing the world, just like you are."

"Well, they lost their Empire," said Mac. "I suppose we have to get used to it as well, and hope to do it with the same sort of grace."

"I don't know," I said. "Putting up with a Mosque at Ground

Zero dedicated to the last conquest of Europe seems pretty damned tolerant, don't you think? Insh'allah."

Mac just smiled and shook his head.

"The greatest Naval leader in Japanese history, killed by a broken code and a high-tech aircraft flown by kids. Amazing," I said, scrawling my name on the credit card receipt. "Great story, Sir."

"Not as interesting as what happened in the wardroom of the USS *South Dakota* when Eddie ran into Terrible Turner."

I looked up with interest. I may have paid the tab, but this conversation didn't appear to be over. "You know, Admiral," I said slowly. "There may be something to be said for being the *last one* to the blackboard, too"

Mac just smiled.

JASPER, MUSH AND MAC

• • •

LEGENDARY SKIPPER OF USS *WAHOO*, LCDR "MUSH" MORTON.

I don't think Lizzie or Meghan knew the details about the spy swap—the one in which a Russian sleeper cell of agents had been rounded up by the FBI, and ultimately exchanged with the Kremlin, just like the bad old days on that Bridge in Berlin.

Didn't matter; I was not privy to the nitty-gritty either, and certainly the Admiral didn't know. He was just back from the

Outer Banks and a traditional summer week with his family at the beach.

The girls were more concerned with the growing anticipation about the final resolution of the Lebron James matter, which is much more important than a suddenly visible component of the secret world. I mean, the continued existence of Cleveland as a city was at stake, you know? It was more serious than someone uprooting the Rock and Roll Hall of Fame and leaving town with it.

I was sitting at the corner of the Willow bar, just a few minutes early for my date with the Admiral. The girls were just getting settled at the end of the bar where old Jim normally holds court, pointedly drinking Bud in the upscale wine bar, and on the whole, I infinitely preferred the company.

Lizzie was a dark-haired beauty with an expectant look, no ring, and Meghan, ring, was a vivacious blonde.

The ladies arrived for a glass of wine, and a light snack from the neighborhood bar menu. As attractive as they were, I understood why Peter paid them special deference. Linen and pearls were the motif, and considering how sweltering hot it was outside, they looked cool and elegant.

It had been a busy day. The Russian spy ring had all pled guilty that morning in the Rocket Docket of Federal Court just down the Little River Turnpike from Big Pink that morning, and the lot of them, ten spies, spouses and kids, and were boarding a Vision charter jet headed for Vienna by the time the Admiral drove over from the Madison in his champagne Jaguar and parked at the curb in front of Willow.

I forget what I was doing, except I seemed to be on the phone a lot trying to find the services of a hydrographic engineer with a Top Secret Clearance and working knowledge of the Empty Quarter of Arabia. I made a note on the office pad to check and see if Vision was a wholly-owed subsidiary of an agency where I used to work or the other side. You never know when you might need a charter with a certain understanding of how things work.

Four alleged intelligence operatives were being processed in Moscow, but they had a shorter flight to Austria, and there was a lesser sense of urgency about it. I thought of my pal Ed, who had been detained by the FSB in some trumped-up charges for nearly a year when the kleptocracy was reasserting itself in the Kremlin, and on the whole, decided I was much better off in the commercial side of the business.

Of course, in this screwy decade, who knows what that is anymore? Except for the Jihadis, I forget who the enemy is. And Sara, the dark-eyed knock-out waitress from a Lebanese family, could make you forget about the threat from the Middle East in a heartbeat.

Then the Admiral appeared beside me, looking tanned and ready after his time at the beach. As we settled in for out interview, the spies were getting on planes, and we were about to talk about the targeting issue for the 313th Bomb Wing and the B-29 campaign against Japan.

But first I pulled a napkin off the stack to start taking notes, and borrowed a pen from the Admiral.

I told Meghan that she was sitting next to one of the people who made the victory in the Pacific possible, and that the Admiral had been part of code-breaking team that made the battle of Midway a winning proposition.

The Admiral leaned over and said that he didn't think the ladies were old enough to know what the battle of Midway was, and Meghan sat up tall and took umbrage.

She certainly did know, and wanted to know precisely how the thing was done. The Admiral told her, and she looked at him with wonder. "Have you ever told that story before?"

He smiled, and his eyes twinkled behind his glasses. "Only about 10,000 times," he said.

We all laughed, and the ladies eventually moved on to do something else while the sun was still shining, and the Admiral and I got down to the business of targeting the Japanese petroleum reserves, and the matter of why he is not entombed with "Mush" Morton and the entire crew of the USS *Wahoo* at the bottom of the La Perouse Strait, the northern entrance to the Sea of Japan.

This is going to take a minute, so you might want to go get a fresh drink.

The girls had been interested in whether we were married, being on the topic themselves, and Mac made quite an impression on Lizzie and Meghan when he announced that we were both eligible, not that they were. But being of the age when everyone seems to be pairing off, they were interested in all the possibilities.

Mac told the story of his beloved Billie, whose given name was Sarah, like the dark-eyed waitress who melts my heart, but who drops the "h." She hovered down the bar beyond Peter as the crowd thickened, causing other males to ooze interest.

Mac told us that Billie's Dad had been committed to the idea that she would be born a boy, and though it did not work out that way, he never stopped calling her "Bill." Her friends softened it to "Billie," and that is what Mac called her all through the marriage, and the long decline that she suffered, starting at the age of 59.

That got Meghan's attention. Mac said he had three careers; one as a Naval Officer, one as a senior member of the Intelligence Community, and twenty years caring for Billie. It was an interesting perspective, given that the girls were just starting their lives as married or about-to-be, and I was dealing with what was happening with the decline of my folks back home in Michigan.

Mac has sailed through all the storms save the last and was bright as a new penny. It was with disappointment that we watched them leave; there was Liz's wedding coming up, and the ladies were focused on the various errands they needed run to make the nuptials perfect.

Once they were gone and our brains could concentrate on something else we got down to the topic I wanted to address, which was his interaction with Major General Curtis "Iron Pants" Lemay.

The cigar-chomping XXth Air Force Commander on Guam was having trouble putting the Empire of the Sun to the torch in the winter of 1944 and the long, hot spring and summer of 1945. Mac had been part of a significant change in target a strategy that helped remove the Japanese sanctuary in the Inland Sea.

Mac told me the CINCPACFLT staff moved incrementally forward in January of the last year of the war.

I have a dozen or more cocktail napkins piled up next to the computer about the bombing campaign, and the hidden story of the 313th Bomb Wing, but it isn't that neat. In fact, it is as messy as the blurred ink on the absorptive paper. We had to pop back to Makalapa Crater, since that is where Jasper Holmes saved Mac's life.

It is a little confusing, since Mac was there during the big war, and then back again as the intel chief during Vietnam, and I was there in the early 1980s. The blur of things and slow change of adamant institutions caused many more napkins to be covered with spidery diagrams.

If you are not in our little tribe, I will have to digress for a minute. Bear with me. Wilfred J. "Jasper" Holmes was an intelligence officer in charge of the Estimates Section of the Fleet Radio Unit Pacific (FRUPAC). The whole enterprise was about code-breaking, and was intensely secret. So secret that a special Army unit was established, controlled directly by the Joint Staff in the Pentagon, to control the distribution of the intelligence derived from it.

Mac was one of the handful of officers on Oahu who interpreted and analyzed intelligence derived from the breaking of the Japanese naval encryption code designated "JN-25," and handled by the ULTRA control system by the Army unit's special security officers.

Holmes was neither a code-breaker-cryptologist, in the term of art—nor an intelligence officer. Jasper was an unrestricted line officer, a submariner, and had commanded S-30 at Pearl Harbor in the 1930s before severe arthritis put him on the beach in medical retirement. Jasper had an engineering degree and joined the faculty of the University of Hawaii. He also was facile with the typewriter, and was a published author in the Saturday Evening Post.

He was called back to active service as tensions rose in the Pacific. His natural aptitude brought him to FRUPAC and the Estimates Section, where his experience as a line officer would be best used to interpret what the code-breakers were producing. The alliance between submarines and spooks has been profound down the years, and most of us still have signed oaths in dusty safes that swear us to secrecy about the depth of it.

The Estimates Section was the first prototype of "all source" fusion intelligence that brought together the best information drawn from code-breakers, spies, aerial reconnaissance and operational reporting to determine the strength, composition and movements of the Japanese military machine as it stretched across the Pacific.

Mac was not an intelligence officer either. He had been commissioned as a deck officer, and was working as Deputy Fleet Intelligence by talent, of course, but in that position by the sheer luck of the draw.

Luck is a funny thing.

There were two elite groups who served as the poster children of glory for the war years. The first were the Fly Boys of both services, bombers and fighters roaring across the wild blue, silk-scarved and goggles against the foe.

The other was the Silent Service of the Pacific war, the submariners who took the fight to the Japanese when everyone else was falling back.

Naturally, Mac chafed a bit at staff duty, and by 1943 was interested in getting in the fight. Well, let me be a bit more precise. The earnest officers at the Hawaii Sea Frontier were periodically inquisitive why a junior Deck Officer appeared to have a cushy shore billet on O'Ahu while the war raged. Mac could not tell them what he was doing, and periodically, Jasper Holmes or Eddie Layton would have to reach out and say "Never mind." Still, the Sea Frontier folks did have a point that Mac did not contest. The way things worked in those days was that the submarine force interviewed prospective officers sending them on a war cruise. Mac put in his request, hoping that a war cruise would put the matter of sea duty to rest, and was approved to join the crew of "Mush" Morton's USS *Wahoo* (SS 238).

FLEET RADIO UNIT PACIFIC, ALSO KNOWN AS STATION HYPO,
THE CODE-BREAKING AND INTELLIGENCE STATION AT PEARL HARBOR.

Mac remembers the bold young skipper well. "He had he biggest hands I have ever seen." Annapolis class of 1930, Morton was a capable and highly aggressive submariner with ice-cold seawater in his veins. Arriving at SUBPAC, he convinced Admiral Charles Lockwood that his next cruise would be to the Inland Sea to remedy the deficiencies of existing torpedoes with the new Mark 18 electric model that was still being de-bugged.

Wahoo departed Pearl Harbor on September 9, 1943, with a mixture of each weapon. She left without Mac. Jasper Holmes decided the day before sailing that impending Japanese operations in the Solomons required the most experienced analysts to be at their desks supporting Estimates, and he turned off Mac's orders.

Jack Griggs was the Wahoo officer that Mac was supposed to replace, and he missed the deployment, too. There was not enough time to find a suitable replacement, and the sub left for the SOJ with her wardroom one officer short.

Mush topped off fuel tanks at Midway Island, and proceeded west to enter the Sea of Japan via the northern neck at the La Perouse Strait around September 20th, and patrolled below the 43rd parallel for about four weeks. The Estimates section at FRUPAC followed his progress in the usual looking-glass manner, based on internal reporting by the Japanese navy.

Wahoo was able to sink four ships in the area, including the 8,000-ton steamer *Konron Maru*, which sank with a loss of 544 souls. The other three ships that Morton sank totaled 5,300 tons.

**Wahoo ON THE BOTTOM AT A DEPTH OF 186FT.
DISCOVERED BY A RUSSIAN CREW IN 2006.**

Mush Morton and his 79 crewmembers were never heard from again. Japanese reporting indicates he tried to break out of the SOJ via the La Perouse on the morning of October 11, 1943. Why he chose to make the run on the surface is unknown, but possibly associated with combat damage.

A coastal artillery battery engaged Wahoo, and later patrol aircraft and surface units piled on as Mush took Wahoo below, trailing an oil slick. A submarine chaser arrived in the vicinity and dropped 16 depth charges. An expanding sheen of diesel fuel two hundred feet wide and three miles long was the last thing anyone saw of Mush the Magnificent and the Wahoo until a Russian hydrographic crew surveyed the wreckage in 2006.

The loss of Wahoo sent shock waves throughout the entire submarine force, and Admiral Lockwood immediately ceased patrol operations in the Inland Sea. The Japanese then had that body of water as an effective sanctuary, at least until Mac had a chance to talk to Iron Pants Lemay on Guam 18 months later.

This is a complicated business, and was clearly going to take another session to unravel. I had a second glass of wine as Mac contemplated having another Virgin Mary, one of the special ones that Peter makes with so many vegetables that it is almost a salad.

"So," I said. "Sixty-six years ago, if Jasper Holmes hadn't stepped on your orders, you would have been bones at the bottom of the La Perouse Strait for seven decades?"

"Yep. Mush got the Navy Cross, posthumous. Jack Griggs and I lived. Luck of the draw."

I was blown away by the revelation and took a sip of a lively white chardonnay that Peter recommended. "Submarines are still relevant today," I said. "Did you hear that three Ohio-class boomers all showed up in Asia last week, appearing at Diego Garcia, Subic, and Yokosuka? They have been modified to carry hundreds of Tomahawk cruise missiles. It was a signal to China, I think, and a warning to the North Koreans over the sinking of that destroyer."

Mac smiled a thin smile. "My sources say the signal didn't work. The big exercise that the Fleet was going to conduct in the Yellow Sea was moved to where the Chinese told us we could operate."

"So the Chinese are telling us where we can go in international waters?"

"Appears to be true. They have established a sort of Red sanctuary, like the Japanese did in the Inland Sea in 1944, I guess."

I shook my head. Where is Iron Pants Lemay when you need him?

USS *Wahoo* Memorial overlooking the La Perouse Strait.

PEAR PIE

• • •

A BAD PIE. MAC STILL HATED THE IDEA SIXTY YEARS LATER.

"So what was it like when the Staff was directed to go forward?" I asked the Admiral. "Leaving Hawaii for the combat zone must have been quite a change."

"Well, yes, of course it was, but Admiral Nimitz wanted to lead from the front. I got Eddie Layton, the Fleet Intelligence Officer, to let me pick the four best analysts at JICPOA and a Yeoman Chief Petty Officer and a First Class and that was the intelligence section at the HQ forward. We got to Guam in January of 1945 and stayed for the duration."

"That isn't much in the way of a staff. Were you the Deputy N2 or the J2?"

"Neither. We were just Fleet Intelligence, Estimates Section. The Navy didn't start with the staff numbering system until after the war. The SeaBees had done a remarkable job is getting things set up. There was a nice Headquarters building on what we called CINCPAC Hill and a mess hall across the street."

"So what did you do? I mean, what was your job like?"

"The first part of the staff day was the joint briefing. That was the one that Iron Pants Lemay from XX Air Force came to at 0900. I put it together based on the material that the Foreign Broadcast Intercept Service copied from unclassified Japanese media from their site on Saipan."

"The briefings were all unclassified?"

CHESTER NIMITZ ON GUAM AT CINCPAC HILL, LATER RE-NAMED IN HIS HONOR.

"Yep, but that was the trick, to find the unclassified information that confirmed the classified material we got from FRUPAC. It was a sort of misdirection to keep the Japanese in the dark and still get high-quality intelligence to people who could use it without all the security bells-and-whistles. I still have copies of all the ones I gave in Guam. Admiral Nimitz did not travel much, but when he did we got him written copies and some of them have his initials on them. Never a comment, but he always signed off. YN1 Harry Truman would take dictation as fast as I could speak the words. He was quite incredible. Of course, even though the briefings were unclassified, they were colored by the ULTRA traffic we got courtesy of Army Captain Chuck Kingston, the Special Security Officer. We also had a direct line to the Estimates Section back at FRUPAC in Hawaii where Jasper Holmes and his staff could do research for us. We used it as a chatter line, too."

"That didn't change," I said. "We still had a teletype line to the other intelligence centers when I was in Hawaii, and I still remember the tall keys and the springy feel to keyboard. The Operators hated us for having a way to talk to the world from the ship that they could not control. Where did you live?"

"I was billeted in a two-story Quonset hut. I shared it with another Lieutenant. It was comfortable. Beyond us were the Flag quarters. Admirals shared some places, and the Captains were there, too. We had a wine mess, too."

"Was there a limit on alcohol?"

Mac shook his head. "No. Let your conscience be your guide was good enough. Plus, the officer's club was on the point beyond that. When we were done with the compound we turned it over to Pan Am. They used it as a sort of R&R facility for lay-overs on the Clipper flights to the Far East once things got rolling again after the war."

"When I married Billie after the war was done, she asked me if there was anything she shouldn't cook. She was pretty good in the kitchen, and after the days on Guam in 1945, I will tell you there was not a great deal I wouldn't eat. I even liked that bulk Spam that cookie would cut up to look like pork chops and bread and fry them. But I drew the line at pear pie. That was my limit, and war or no war, I stuck to it."

C-RATS

I glanced at the delicate pork spring rolls that came out of Tracy's kitchen on the $5 neighborhood bar menu. I love those things, and with the spicy dipping sauce, treat them like a meal.

"In Hawaii, Spam was a delicacy. So were those canned sausages-what were they?"

"Vienna sausage. They used to serve them with chili, too. Anything with enough hot sauce to kill the taste. Nothing out forward could beat those Local plate lunches we could get down on Hotel Street in Honolulu. Red hot dogs with chili and 2 scoops of mac sal."

"When we lived there, I used to fry up slices of Spam and serve with eggs and hot sauce," I said. "That was when we were on shift work at PACFLT and the Soviets had ballistic missile subs continuously on patrol in EASTPAC."

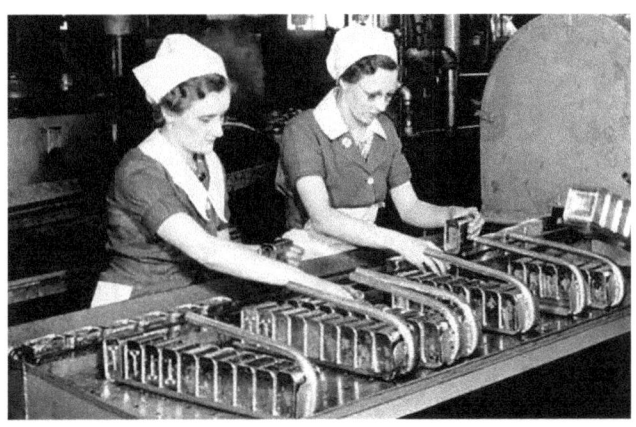

WARTIME PRODUCTION OF CANNED SPAM.

"Living on an island can yield some interesting culinary quirks," said the Admiral, taking a sip of his Virgin Mary. I dipped a spring roll in the dark spicy sauce.

"Macaroni salad is usually just elbow macaroni and a heaping portion of mayonnaise, just like in the South."

"On Guam, there wasn't any mac sal. The CINCPAC Forward HQ dined on C-rats, exclusively."

"What was it like?" I asked, thinking about eating out of cans for eight months.

"Well, specifically, the Type C ration was part of family of food types for the forces in the field. A-rations were fresh food. B-rations were packaged un-cooked food. C-rats were canned food that came in a monotonous variety of flavors. Meat and beans, vegetable stew, meat and spaghetti, ham-egg-and-potato. Pork and rice, franks and beans, which were just like Vienna sausage, and that awful ham and lima beans."

"So when bulk food came in it was a big deal?"

"Oh, you bet. The problem was the fruit. Cookie didn't have much in the way of desserts, so he settled on those cans of pears to fill his crusts of lard and flour." Mac frowned. "They were as heavy as man-hole covers. He would make a pie out of those pears, and even sixty-five years later I can't stomach the thought of it."

I was going to ask the Admiral about how his unclassified briefings caused Iron Pants to mutiny against the Joint Target Board back in Washington and assign the 313th Bomber Wing to the Navy's idea of targets, but I was having a hard time getting Peter's attention to get another glass of chardonnay. The very thought of fried Vienna sausage with a side of *mac sal* was making me hungry.

UNDER CONSTRUCTION

• • •

I knew I was in for it when Mac's gold Jaguar saloon pulled up in front of my favorite kind of saloon. I don't know how he does it—but though he is only driving from the below ground parking across the street, he prefers not to tempt fate waking across busy (insane) Fairfax Drive traffic. He always seems to get Rock Star parking, which is only his due. I was at my usual seat next to Old Jim, but I stood as the Admiral walked slowly into the bar. He was carrying a manila folder.

"I need to clarify a couple of subjects in your understanding of the way things were located in the Pearl Harbor complex during the war years" "He ordered a Virgin Mary from Boomer, with an array of garnish that made it more meal than a drink. Her brash energy behind the bar is contagious. It is Mac's current favorite of the non-alcoholic beverage family his doctor prescribes.

Mac pulled out what looked like an old-fashioned aeronautical area photo.

"I'll explain these, and you can put them into your own words for whatever you do with them after these interviews."

"First," he said. "The location of HYPO prior to its move to the two-story building in Makalapa sometime in the summer or fall of 1943:

"Prior to the beginning of the war, the CINCPACFLT headquarters were located on the second deck of a building on the Submarine Base, and it remained there through at least most of 1942 while the new building in was under construction."

"The Combat Intelligence Unit (CIIU (or HYPO) was located in the 14th ND Admin Building in the Navy Yard. After Rochefort arrived in June 1941, he insisted on non-visible and more secure spaces for the CIU, which had been co-located with 14ND operations on the second deck of the Admin Bldg. "

"The result was a move to the "unfinished" basement area of the Admin Bldg. That area had been a non-ventilated storage area, so it needed some work to convert part of it to office space where people could work. Primitive air conditioning equipment was installed, but it only cooled and did not de-humidify; hence, "smoking jackets" and sweaters were needed by most of the workers. That is where all those rumors about Joe dressing eccentrically. We called the place The Dungeon."

"I have been down the stairs to the front door of the place," I said, "but the nuke submariners use it for some kind of training and normally won't let anyone in."

"Not much to see now," he mused. "HYPO remained in the basement spaces until the late summer or fall of 1943 while its new Makalapa building was under construction."

"Late in 1942, ICPOA was formed. Rochefort was the first Acting OinC of ICPOA, but their offices were put into the adjacent supply building. Both the CIU (HYPO) and ICPOA were administratively part of the 14th ND because ADM Nimitz wanted to keep his immediate staff as small as possible. Thus, throughout the war both FRUPAC and JICPOA (totaling eventually near 5,000 persons). although located in Makalapa buildings, remained administratively subordinate to COM14 (equipping, housing, feeding, etc).

"Any confusion on who the customer was for your product?" I smiled and took a sip of chilled white.

"But, let there be no mistake; we were there to support CPF and his subordinate operating commands. Our administrative subordination posed no interference to that mission. (The beginning of the basement occupation is discussed in Chapter 7 of the Rochefort book, and there is a diagram in the first set of photos.)

"So, at the time of Midway, CPF was still at the Sub Base, the Makalapa headquarters was under construction, and the CIU was in the basement of the 14ND Admin Bldg. The only event I can associate with our move to Makalapa was the shoot-down of ADM Yamamoto on 18 April 1943. We were still in the Navy Yard basement at that "memorable" time."

RENDEZVOUS

● ● ●

"I have often said that an intelligence officer has one task, one job, one mission. This is to tell his commander, his superior, today, what the [enemies] are going to do tomorrow. This is his job. If he doesn't do this, then he's failed." – Captain Joseph J. Rochefort, USN

I was having one of Willow's specials off the Lunch Counter menu—the turkey matzo ball stew—and talking to Brian about growing up in Honolulu. He brought a couple ancient yellowed articles with him, papers his Dad the Chief had saved to recall the events of people he knew during the war.

One of them was about the second attack on Pearl Harbor, the one that was a military secret at the time, and was largely lost to history for years. I understand the reasons at the time.

No point in getting everyone worked up again, and from a military perspective, once they figured it out, the staging base for the seaplanes at French Frigate Shoals was patrolled and denied to the Japanese. That also might have contributed to the lack of long-range patrol aircraft to look for the American fleet before the battle of Midway.

I dealt with all sorts of secrets in my Navy years—some large and some small. Some came with a fifty-year seal on them, and my pal The Good Doctor put his history of the program he supported in the safe with a sigh, knowing no one would ever see the stories contained within.

There are still some things about Pearl Harbor that are under seal—and one can only wonder what will pop out when the 75-year window of secrecy expires in 2020.

We looked at the yellowing article in the Honolulu Advertiser. I remember the first time I heard of the second Japanese attack like it was yesterday.

Mac was 91 that year. He drove over to the Willow in the champagne-colored jaguar sedan from his residence at the Madison across the street and got his million-dollar parking space out front. I looked over at him in his aloha shirt. "So tell me how it all went down, Sir, just for the record. You have told the story a thousand times, but I want to bounce it off what Jasper and Eddie wrote about it later, and the oral history Joe Rochefort did in 1969. You are the only one who can put it all in context now."

I was feeling expansive, pleasantly lit up with Willow's current vintage of an insouciantly dry Spanish white that Big Jim the bartender would top off periodically. I felt that we had beaten the year of 1942 about to death. We had talked about Midway, and the growing intelligence organization, and the treachery of the Washington Radio Intelligence Mafia (Wenger and the Redmans), and I was eager to get on to the Spring of 1943, since I was getting desperate to kill Admiral Yamamoto again.

"You know I went down to Honolulu on Christmas Day of 1942 and went through the barbed wire for a swim from the beach," said Mac absently. "Then I wrote to the folks back in Iowa and told them about it. They would have been freezing then."

American kids were dying in the jungles of the Solomon Islands. Hundreds of other kids had been blown to bits or drowned in The Slot in the see-saw battle for Guadalcanal as the Japanese ran the Tokyo Express in at night to bring reinforcements to the islands, and the Americans flew from Henderson Field by day. Back in Pearl, and Melbourne, Australia, the code-breakers labored eighteen and twenty hours a day to recover the values of the new five-group system imposed by the IJN.

The Admiral took a sip of his Virgin Mary and waved a colossal olive on the end of a toothpick at me. "They don't talk about the second attack much, do they?"

I sighed. This interview was not going the way I had hoped. "What second attack? I know the Japs had used balloons to send incendiary devices to the Pacific Northwest to start forest fires, and they had submarine-carried seaplanes to attack the Panama Canal, but that didn't happen."

"Well, the second attack on Pearl Harbor did. It was on the 4th of March, 1942. Eddie Layton did some research after the war and discovered what had gone down, and it might have been Jasper's fault."

"What on earth do you mean," I asked. "How could Jasper have helped the Japanese?"

"Well," said Mac. "You know that Jasper was a pretty successful author. He wrote for magazines like the Saturday Evening Post under the pen-name Alec Hudson."

"Yeah, I know. I ordered a copy of "Up Periscope" the other day, and just started it. He writes well."

"I have that one and his "Doubled-edged Secrets, both auto-graphed," said Mac. He smiled and reached down to his briefcase and produced a book with a black cover, the title superimposed in lines of gray. "This is the one I gave to my mother, and I got it back after she passed." He opened it and showed me the inscription:

"To Hattie Showers Whose son had a very important part in these
events. With the compliments of the author.

W. J. Holmes"

"So what about the second attack," I said after looking at the words. "I have never heard about that and I lived and worked in Pearl and thought I knew everything about it."

Mac smiled. "It was two months after I arrived in the basement of Building One. Joe Rochefort was anxious about the possibility that the Japs might come back, and he was right. There was an article about it in Proceedings a few years ago, based on some research that Eddie Layton did in the 1950s, and a series of interviews that Joe Rochefort did in 1969. It was embarrassing, and that might be the reason people don't talk about it much."

"What happened?" I said, mystified.

"The Japanese decided to mount a follow-on attack against the shipyard at Pearl to destroy the big Ten-Ten Dry-dock. That would delay the repairs to the battleships, and increase the paranoia on the island. They envisioned an attack by five big Kawanishi H8K "Emily" class flying boats.

"Jesus," I said. "And they pulled it off? Why isn't that part of the big history?"

"Well, that is the interesting part. Eddie located the Japanese OpOrders and the pilot reports of the mission after the war. As it turned out, only two seaplanes departed the Marshalls, and they did refuel from a submarine milk cow in the vicinity of French Frigate Shoals.

They then launched for Oahu, to bomb and conduct what Joe Rochefort described as an 'armed reconnaissance' mission."

"And there was no response, only four months after the biggest disaster in American military history? No warning?"

"Oh, there was warning all right. We had penetrated JN-25 enough at that point to know that something was up, but there was heavy overcast the night of the 4th of March and the Japanese got lost. The two Japanese planes wandered around over the island but the blackout was effective, and one of the planes dropped bombs on Tantulus, which produced a couple large craters and broke some windows at Roosevelt High School. Eddie thought the other plane must have dropped its bombs over the ocean."

"That was it?" I asked. My wineglass stood empty next to the notepad. My pen was making exclamation points next to the words "Second Attack on Pearl: no response!!!!!"

"The submarine I-23, the weather reporting unit, disappeared while on patrol and never arrived on station. So the weather was bad, the planes never found Pearl Harbor, and there were two explosions in the night that the next day the Navy and the Army blamed on each other for dumping ordnance irresponsibly."

"Good God, that would have been hugely embarrassing if word got out that the Japanese came back and were not intercepted."

"Joe Rochefort said it this way," Mac said, pulling a folded article out of the back of his copy of Double Edged Secrets. He unfolded it and peered over his glasses. "He had passed a warning to the 14th Naval District, the Hawaiian Sea Frontier Commander, and to CINCPAC. This is from the interview he did with Commander Ette-Belle Kitchen in 1969."

He began to read in a voice just loud enough to be heard over the dignified din in the Willow bar:

"I was told later by informed people that the attack was made, as I say, more or less unmolested, because the Navy had no airplanes at that time capable of repelling this attack or destroying the incoming aircraft. The Army said that they only had one-place fighters, and who could expect a fighter pilot to not only fly the plane in darkness but also to approach and make an attack on any enemy plane. Therefore, nothing had been done about it, and no action was taken."

"My God." Mac smiled at me and handed the paper over. I looked at the rest of what Joe Rochefort felt then:

"I just threw up my hands and said it might be a good idea to remind everybody concerned that this nation was at war.... It's not a very glorious incident. You won't find very many references to this anywhere along the line."

I re-folded the article and slipped it back in the book.

"That certainly beats me," I said. "So that was the end of it?"

"No," said Mac. "They planned on trying it again before Midway, but Admiral Nimitz dispatched a couple seaplane tenders to French Frigate Shoals, and the Japanese wouldn't risk it. That meant they lost a chance to see what we were doing before the Midway battle in June."

"Amazing," I said. "That is just one of the reasons I enjoy talking to you so much."

"There was a sort of gallows humor about it on the CinCPac staff," said Mac. "You will see that one of Jasper's stories in "Up Periscope" is called "Rendezvous." It was originally in the Saturday Evening Post just before the war. It outlines a plan by which submarines would refuel long-distance seaplanes for a sneak attack. There was quite a laugh about it on the staff, at least those that were cleared. Eddie Layton suggested that Jasper had planned the attack for the

Japs, but it was not completely in jest. There was an investigation, and Jasper was exonerated."

"That must have been sort of strange between pals," I said.

"They weren't," said Mac. "Joe Rochefort and Eddie were very close. But Eddie couldn't stand Jasper. I was one of his kids in the Estimates Section, and he was delighted when he finally got enough of us Lieutenants to stand up a real 24-hour watch. And he made Commander."

"I imagine that would pick up anyone's spirits," I said. "It certainly did for me."

"After I volunteered to go forward to Guam in '45, Jasper came out to make a visit. I told Eddie I was going to go out to the airfield and pick him up in my jeep. Eddie wanted to know why I was going to bother."

I shook my head in wonder at the memory of a rendezvous so long ago and far away, vivid as if it were just yesterday.

It is a comfort to duck back into Mac Shower's account of life at the forward headquarters, and how he and a mad Air Corps General cooked up a private plan to destroy the wartime economy of Japan. Things used to be a lot simpler.

IRON PANTS AND CHERRY PIE

• • •

I took a sip of chardonnay. Peter was pouring a very nice vintage, a hint of fruit but dry, without anything that seemed syrupy like canned pears.

I looked over at the Admiral in wonder. He had just described the entire intelligence staff that went forward with Fleet Admiral Chester Nimitz. Captain Eddie Layton, Lieutenant Mac, four analysts from JICPOA, a Senior Chief Yeoman and First Class Petty Officer Harry Truman.

"So let me get this straight. You had a grand total of eight guys supporting a five-star staff that controlled a war effort of 2000 ships, 25,000 aircraft and 2.5 million men?"

The Admiral nodded. "Don't forget, we had a secure telephone to talk to Jasper Holmes and his Estimates Section at FRUPAC."

I was stunned that anything so complex could be accomplished without battalions of analysts and flat-screen panels. I had seen a request for hundreds of bodies to go forward to support operations in Afghanistan the other day, not fighters, mind you, just thinkers of great thoughts.

They must have been giants then, or rather, ordinary men who rose to extraordinary heights because there was no alternative.

Mac did the daily brief to Chester Nimitz's staff on CINCPAC Hill above the harbor at Agana, Guam. He pulled the neat trick of giving it at the unclassified level, which enabled Captain Layton to spin the message to the broadest possible audience. The unclassified message was fitted to the template of things that were known to be true from highly classified ULTRA sources and exploitation of the Navy and Army tactical codes.

It was pretty slick, and I have done that myself to be able to share information with third-party allies in times of crisis and conflict. 'Why' you know something to be true or not is not the point; sources and methods do not need to be revealed if the information is correct.

I had learned the lesson just as clearly long ago. One evening in the Yellow Sea I had labored long through the night to plot the intricacies of an exercise scenario that I had to present to the crews of the dawn launch from *Midway Maru*.

Two hours prior to commencement of flight operations I labored through the complexity of the scenario to the bleary men in green NOMEX flight suits sipping coffee and puffing on their second or third cigarette of the morning. I got through it and Santa, a lanky Radar Intercept Officer ('RIO') from the Rock Rivers Phantom Squadron, snarled at me: "Just tell me where we are supposed to be and what we are supposed to do when we get there?"

I got it, suddenly and completely. Don't explain how to build a watch when someone asks what time it is.

One guy who did not need to know about watch construction was Iron Pants Curtis Lemay. He was in the audience when Mac was briefing the staff at 0900 each day. Iron Pants commanded the XXth Air Force. It was the first such command not put under Theater command of officers like Eisenhower, McArthur or Nimitz. Instead General Curtis LeMay assumed direct command of the 20th Air Force. The mission of the 20th was to engage in the strategic destruction of Japan by air, with Iron Pants providing the direction under the guidance of Washington. Which wasn't working for him. LeMay wanted to cut the legs out from under the Japanese economy, not just aircraft engine production.

Long afterward, that Whiz Kid Robert McNamara described LeMay's approach to addressing the abort rate by bomber crews in the European theater. The crews were no fools, and they exercised a certain discretion on "down gripes" on their birds that might enable them to return to home base and keep them out of meat-grinders like the raid on Schweinfort.

MacNamara noted: "He was extraordinarily belligerent, many thought brutal. He got the report about the aborts. He issued an order. He said, 'I will be in the lead plane on every mission. Any plane that takes off will go over the target, or the crew will be court-martialed.' The abort rate dropped overnight. Now *that's* the kind of commander he was."

Iron Pants listened to what Mac said in the morning brief in Guam. Now commanding the 20th Air Force and hundreds of sleek Super Forts, he had been directed to improve the efficiency of the bombing campaign against the Japanese.

Eddie Layton and Mac knew that it was not the destruction of aircraft engines and ball bearings that would shut down the Japanese war machine. It would be a lack of oil and lubricants. The unclassified reports of distilling the roots of pines to make fuel were buttressed by highly classified intercepts of military communications. Eddie Layton and Mac knew the truth.

Iron Pants Lemay decided to ignore the idiots back in Washington, and gave an entire bomb group to the Navy to shut down the Inland Sea.

Mac agreed to tell me more about the 313th Bomb Wing, but I was busy. I was conducting an experiment in World War II cooking. I was attempting to deconstruct Mac's description of the hated recipe for Canned Pear Pie a la Cookie last night. I could not make a crust work with the crackers that used to come with C-Rations, even with laden with lard and smoothed with canned milk.

It was like Mac told me: "Canned pears are notoriously flavorless, grainy, and colorless. UGGGHHH. Just unappetizing & tasteless. Now for cherry pie. . . with real crust and the sweet tangy goodness of the cherries. But we had to get back home to experience that taste again."

OPERATION STARVATION

• • •

I nodded as Mac told me of the revolting matter of Pear Pie, made out of the grainy fruit contained in sugary syrup in gigantic cans and crushed crackers as a revolting crust.

I was having the neighborhood bar menu Spring Rolls with dipping sauce, and was considering getting an order of the mini Fish and Chips, which is not what you would think as prepared at the Willow Restaurant. Tracy O'Grady's kitchen team floats out a little fantasy plate that is more akin to the finest Japanese tempura, garnished with slender rings of onion in the same delicate batter.

"There is nothing that sailors care more about than their chow,"

I said, after Mac described the constant searching for something decent to eat on Guam in 1945. When afloat, in war or peace, there is nothing to do except work, sleep or eat. Consequently, what Cookie manages to get on the mess line or down to the Dirty Shirt Wardroom is the only thing that marks the passing of the hours in the endless sameness of the ship's routine.

Bad food is bad morale. I remember vividly the stories of the Peanut Butter riots on the Coral Sea during the Vietnam conflict, when the wardroom treasurer ran short of funds, and the brown paste was the only thing for lunch for weeks.

Ashore, as Mac was on Guam, there was no adequate means of transporting fresh food, and the Staff forward was reduced to the dreary and numbing sameness of canned C-rations.

Mac said the driving from the Headquarters complex on CINCPAC Hill on Guam helped to vary the boredom. They could watch the strikes launch from Anderson Field in the morning and grab breakfast at the Air Corps mess when the tempo of operations permitted. The flyboys were able to get fresh food via cargo planes from Hawaii and they had things like real butter. Their mess was a treat, a taste of home in the forward area.

The ring was closing on Japan, and that is why Chester Nimitz took his command element forward. The main event, the invasion of the Japanese Home Islands was going to come in this pivotal year. There were other options, of course, but in January there was no certainly that a wonder weapon would appear that would change the course of history.

For the foreseeable future, it was going to be B-29 Super Forts in endless waves, putting the torch to the enemy from the new airbases in the Marianas.

Everyone had moved forward to join the fight. The Joint Radio Analysis Group, Forward Area (RAGFOR), set up shop as soon as the shooting died down in September of 1944.

The CINCPAC Staff began preparations to join them shortly thereafter.

In August 1944, Iron Pants LeMay transferred to the China-Burma-India theater and directed first the XX Bomber Command in China. In January, the disappointing results of Brigadier General Haywood Hansell's high-altitude precision campaign resulted in iron Pants being transferred to relieve him in January, the same time Mac and the little Fleet Intelligence organization set up shop on CINCPAC Hill.

Iron Pants would henceforth command the XXI Bomber Command. In that role he was responsible for all strategic air operations against the Japanese home islands. Consequently, he attended the 0900 staff briefing at the CINCPAC HQ.

LT Mac scanned the classified traffic generated by RAGFOR and FRUPAC in Hawaii to which they had a direct line. Armed with the most sensitive communications information, he was able to shape the reporting collected from Japanese public radio stations to form a perfectly unclassified and uncannily accurate assessment of the conditions in Tokyo.

The shortage of petroleum products (POL) revealed in public media reports of was one of the key issues that Mac hammered home under the guidance of Fleet Intelligence Eddie Layton.

Lemay heard each morning that the targets his aircrews were hitting were the wrong ones to put the squeeze directly on the Japanese war machine.

Mac would dictate the Foreign Broadcast Intercept reports that fit the all-source intelligence assessment to Yeoman First Class Harry Truman who typed up the notes as fast as the words were spoken.

Besides POL, the key issue for the Japanese was food, a large percentage of which arrived by ship across the Inland Sea, or from fishing grounds in the Sea of Okhotsk.

LeMay knew he had a hell of a problem. It was readily apparent that the tactics developed for use in Europe against the Luftwaffe were unsuitable against Japan. His bombers flying from China were dropping their bombs near their targets only 5% of the time. Operational losses of aircraft and crews were high, due to Japanese air defenses and the continuing mechanical problems with the B-29 engines.

In Washington, the Air Corps leadership viewed anything except high explosives against land targets as "the Navy's job." Even today, the aviators of both services still enjoy "visually pleasing destruction" as the direct feedback of a dangerous job well done.

I am sympathetic to that view, but it was not the answer to bringing down Japan.

Mac said that as an island nation, the Japanese relied on imports of nearly everything, including 80% of its oil, 90% of its iron ore, and food of course.

PACFLT submarines has carried the burden of attacking the

Japanese merchant fleet since 1943, but with the loss of USS *Wahoo*, no offensive operations had been permitted by SUBPAC in the Inland Sea. It was an open road from the Asian mainland to the west coast ports.

Bolstered by the assessments provided by Captain Layton and Mac, Chester Nimitz decided to push Japan over the edge. That meant an effective and complete blockade, but he needed Lemay's Super Forts to do it. He intended to use the bombers to deliver not bombs, but sea mines in the strategic choak-points that entered the inland sea. Submarines would continue to shut down the east coast ports.

Even the pointy-headed targeteers of the Committee of Operations Analysts in Washington took the Fleet Admiral's request seriously. They, too, were beginning to understand the assessment from Guam that the only way to shut down the Home Islands was an effective and complete blockade.

Iron Pants was a stubborn son-of-a-bitch, and was reluctant to give up control of any of his bombers. But Mac's briefings convinced him of the merit of mine-warfare. He agreed to devote the 313th bomb wing to strike what was euphemistically called "local target-ing," which was a way to say "naval targets" without acknowledging that the Fleet Intelligence people were calling the shots.

The B-29 aerial mining campaign began in late March of 1945.

Mac dipped a spring roll in the dipping sauce and downed it in two bites. "I never heard anything about this, Admiral," I said in wonder.

"You never will. The Air Force historians are not interested in the very brave performance of Air Corps crews that doesn't support the notion of an independent service. I have offered to tell them the story, but got no takers."

"Did the operation have a name?" I asked, looking back at the menu. "Yep," he said quietly. "We called it Operation Starvation."

ERNIE (AND MAC'S) WAR

• • •

WWI SAILOR AND WWII WAR CORRESPONDENT ERNIE PYLE ON THE USS *CABOT*
WHEN HE WAS MAKING WAVES WITH THE NAVY DEPARTMENT. MAC MET HIM
ON GUAM WITH THE CINCPAC FORWARD STAFF. PHOTO: U.S. NAVY

This Spring Ahead nonsense with the clocks is kicking my butt. The computer is telling me it is past noon—and that is nonsense. Why do we not have the courage to leave the sun alone? Who is it in Congress that believe themselves as King Canute, waving not at the ocean in this case, but the hurtling blazing orb of Old Sol?

Vanity, vanity, thy name is Congress. Or something. I felt jet-lagged right at the dinner table where the laptop lives.

I tried to look at my notes from the last session with Mac at Willow and they are not making much sense. Topically, the conversation veered from:

The delicious Gruyere Cheese puffs, and the crackling Peking Duck pancakes, a nurses report from INOVA Fairfax that revealed nothing wrong with Mac, 50 miles covered in the champagne Jaguar, gall bladders, fried chicken done to perfection early in the last century, the obits of two men I did not know, the status of traffic on I-66 eastbound, and what the six fire trucks and twenty-five police cruisers were up to, Section 66 of the Arlington National Cemetery, and in-ground placement of urns therein, the deficiencies of the original Columbarium at the National Cemetery, the status of bartender Katia's job offer from outside the food, beverage and hospitality industry; whether the consolidation of the weather guessers, Cyrppies, Public Affairs and Intelligence Officers in the Navy Corps of Information Dominance was 'back to the future,' since Mac had started as a special investigator who did PAO stuff at the Naval District in Seattle in 1941.

I was not making much progress on my happy hour white, since we were all over the map. "So, you were really a Public Affairs officer before you were a codebreaker, right?"

Mac nodded. "It was part of the Office of Naval Intelligence," he said. "I guess they thought it was all information and pretty much the same thing."

"Maybe they do again," I said. I finally asked Mac about Chester Nimitz, the phlegmatic Texan who led the Navy's drive west across the Pacific.

Mac was much more focused on this, and it was a little unusual, since the mythic figure of the Fleet Admiral normally was part of the backdrop to *his* war.

Mac furrowed his brow, attempting to distill the legend from the man. "Well," he said slowly. "He took care of his Enlisted guys—the ones in the motor pool and the boat detail for the Flag Barge." He went on to describe his conduit to the troops, who was a Mustang Lieutenant who had enormous impact on the staff and the way it worked.

I made a note to find a roster of the staff from 1945, and see if I could track down who the officer had been, since he showed up again as an agent of influence who secured the Junior Officer BOQ across from the HQ after the war—where the chapel is now. It was a rare name that Mac did not remember, or did after I was scribbling something else. I made a note to check it out and ask more questions.

Mac was still contemplating the Public Affairs question. "It was interesting to see who came forward to Guam. Like Ernie Pyle, the legendary war correspondent."

"He was the most famous correspondent of the War—probably more than Edward R. Murrow. Did you ever meet him?"

"Oh yes. He was on Guam before the invasion. Ernie was shot by a Japanese machine gunner on Ie Shima, Okinawa."

"There is quite a display for him in the Hall of Correspondents in the Pentagon, and I would stop and read the panel display when I spent more time than I wanted walking to and from press conferences with The Joint Staff. Ernie was known as the soldier's correspondent, according to the display. Did you ever have drinks with him?"

That is the way of these conversations—I had no inkling

whatsoever that in the course of this session we would stumble across the most iconic and tragic media figures of the war in Europe and the Pacific, two theaters whose paths seemed rarely to cross.

"Not true," said Mac. "Don't forget Admiral Sir Bruce Fraser, F-R-A-S-E-R. He came to the Pacific to command all British naval forces after commanding the group that sank the Nazi battleship *Scharnhorst* off Norway."

"Point taken, Sir," I said, and that led to a discussion of the wild four-day party that started on board Cagey Five, the battleship HMS *King George V.*

"But you actually met Ernie Pyle? That is incredible."

The Admiral shrugged. "He was there, it was a small island. He had sort of prickly relations with the Navy, since he thought the sailors had it pretty cushy compared to the combat infantry of the ETO. He was a plain-spoken SOB and he wore his heart on his sleeve."

"He might not have thought that if he saw the results of a running gunfight like the battle for the Slot," I said indignantly. "I just read Neptune's Fury, Jim Hornfischer's account of the slaughter."

Mac looked on with the cool perspective that only ten decades on the planet can give you, and know you are the last man standing from all the formations so long ago. "It is one thing to read about it and quite another to live it. Ernie had a point. He got bombed with our guys by the Army Air Corps at St. Lo, and was badly shaken. His heart was always with the riflemen of that war. In 1944, he wrote a column urging that soldiers in combat get "fight pay" just as airmen were paid "flight pay." Congress passed a law authorizing

$10 a month extra pay for combat infantrymen. The legislation was called 'The Ernie Pyle Bill.' *That* was the mark of the depth of fondness the troops had for him."

"Sounds like Bill Mauldin and his Willie and Joe cartoons."

"Close enough," said Mac. "And you can throw Andy Rooney in that group, too. Andy was one of the angry young men, then. Ernie was an old man, though—he was 45 when he was shot. I remember thinking about that at the time. I was still in my mid-twenties and he was old enough to be my father."

"Do you recall how you heard about his death? Did you have to clear the dispatches or the pictures before they went back to CONUS?"

Mac shook his head. "Not that I recall. There *was* a picture I might have seen at the time, but it never was published. I think it was the middle of April in '45. He went from Guam to Okinawa to cover the action. These days, you would say he was 'imbedded' with the 77th Infantry Division."

"He took all the risks that the grunts did," I said. "That is a commitment to the mission."

"Yes, I think so. He was riding in a jeep with the CO on an infantry regiment and a couple other guys. Apparently, hundreds of vehicles had driven the same road, but for whatever reason, a Japanese machine gun opened up on them. They stopped the jeep and everyone jumped into the ditch. Apparently, Ernie raised his head to ask the Colonel if he was OK."

"I think I heard he was killed by a sniper," I said, stopping my scribbling.

"Nope, it was a Jap machine gun that had played possum, letting hundreds of other vehicles go by. Those were Ernie's last words, though. He took a round in the temple and was killed instantly right after he asked."

"Maybe that is why the machine gunners waited since they must have known that opening up would get them killed pretty quickly. That is a powerful argument for keeping your head down," I said.

"Yes indeed. I am sure Ernie would have preferred for it to work out differently. They Army buried him with his helmet on with a bunch of the other combat dead. He was one of the few civilians to be awarded the Purple Heart."

"Wait, I saw his grave at the Punchbowl in Honolulu!" I said.

"Ernie traveled a while after the war. They exhumed him from the grave in Ie Shima, and then buried him in the Army cemetery on Okinawa, and then finally they moved him to the National Memorial Cemetery of the Pacific where you saw him."

"That is amazing," I said. "He was the most-read correspondent of the war."

Mac nodded and finished his Bell's Lager. "He even impacted the Japanese. When Okinawa was returned to Japan's control after the war, Ernie's monument was one of only three American memorials allowed to remain in place."

"Huh." I fished around in my wallet, and the Admiral picked up the tab, which was modest since he drinks for free now at Willow.

We settled up with Liz-S and Jasper, the best bowler on Guam, and went to find the Admiral's walker. It is a high-speed model, with hand-breaks brakes and quick action movement. I walked back across Fairfax Drive with him until he turned off at the entrance to the Madison. I retrieved the Bluesmobile from the Bat Cave under the hotel and drove home.

Once I opened the mail, I poured a tall one and logged onto the 'net. I checked out the Ernie Pyle Monument, in place for 67 years.

HERESY

• • •

Mac took a sip of his Virgin Mary and scowled. "I think there are only two olives in this one." He peered into the dense red of the contents of the pint glass before him, rheumy blue eyes squinting behind his silver-framed glasses.

The long bar was filling up nicely. Old Jim was in his position at the apex of the Amen Corner, slowly and deliberately wrapping the cord of his media player around the small electronic rectangle. At precisely 5:15 pm, Peter dialed down the lighting in the Willow bar to increase the romantic ambiance with the rich dark wood and the little votive candles that suddenly increased their bright glitter along the long bar.

Andre-the-Waiter, phlegmatic and cool with his shaved head and impressively articulated physique, was most solicitous. He brought me the mildly insouciant white that Peter was flogging at happy hour prices without need of my beckoning. We continued the conversation about tightening the belt on Japan.

"Two things you need to understand. Iron Pants LeMay came up with incendiary bombing of Japan as a tactic on his own. He took over command of the strategic air campaign in January when we got there. High altitude precision bombing was an oxymoron in the weather conditions over Japan."

"Yeah," I nodded, scribbling a note on a napkin. "It was usually cloudy when I lived in Yokosuka. I had the Flight Deck Integrity Watch one morning when it wasn't, and I realized if it was clear, you could see Mt. Fuji from the carrier pier, looming as an invisible presence most of the time. It wasn't often you could see it, though."

"Precisely. LeMay considered all the options. The climb up to 30,000 feet caused the Super Forts to burn so much fuel that their bomb loads were reduced in order to carry additional fuel tanks in the bomb bays. Plus, the engines were fragile, and less stress on them meant less maintenance. Iron Pants committed heresy, violating doctrine, and he had his crews train at low-level delivery, sometimes down only 5,000 feet. He also decided the only way to bring up accuracy was to transition away from iron bombs to incendiary devices."

"That seems horrific," I said. "It was a distinct change in approach. I read that Air Corps General Ira Eaker once said that the strategy was to kill skilled defense workers. Those are civilian targets."

"It was horrific all around. The early incendiary devices were unstable, and one went off during unloading on the hard-stand and killed a bunch of ordnance men and wounded dozens. The Super Fort was a write-off. As to whether it was moral or not, the first firestorms were visited on the Germans, without any mercy. Same tactic. The Japanese cities just burned better."

"Didn't they start bombing at night, too, like the Brits in Europe?"

"Yes. Japanese air defenses made daytime bombing below the jet stream altitudes very dangerous. LeMay finally switched to low-altitude nighttime incendiary attacks as his bread-and-butter tactic, with daylight high-altitude strikes reserved only for special targets in clear air mass. The first big night fire raid went against Tokyo on the ninth of March, 1945."

"You said there were two things I had to understand. What was the other one?"

"Eddie Layton said the Joint Target Board back in Washington couldn't tell a warehouse from a whorehouse. We kept telling Admiral Nimitz in the morning brief that Petroleum-Oil-Lubricants (POL) were the key to ending the war. LeMay became a believer, and he asked Washington for permission to start attacking POL-related targets."

Mac scowled. "They said: "No, we know best, and they continued to direct the target list against the things they thought were important, like industrial plants. KT. Johnson was one of the big-wigs back in the JTB, and he is still around. I hear from him once in a while.

Anyway, Iron Pants asked for POL targets, but K.T. wouldn't give him any. That is when he decided to commit another act of heresy and gave the 313th bomb group to the Navy."

"That was the "local targeting" euphemism, right?" I said, scribbling away.

"Yep. The Ops guys were getting the idea that the Super Forts could be used to deliver aerial mines and seal up the Inland Sea, and then mine the harbors. Tighten the belt on them and starve them out. LeMay was reluctant at first, but he went along and the first mining missions were flown in late March. After that, the Shimonoseki Strait was effectively closed, and then Henashi Cape, Iwase and Seishin.

The Admiral looked off across the crowded bar, the lights of the votive candles reflecting off his glasses. He recited a litany in sing-song Japanese. "Oyama, Niigata, Miyazu, Maizuru, Tsuruga, Nezugaseki, Obama and Kobe-Osaka. They were mined and re-seeded as necessary by July. The Japs were being cut off."

"No one knew about the Bomb, right? It must have come as quite a surprise."

"That's right. We were on Guam to manage the invasion of Kyushu, the southernmost of the Home Islands. That was Operation Olympic, which was put back a little due to the controversy about casualties."

"When was that supposed to happen?" I asked. "November of '45."

"What was the controversy?"

"We said it might take 2.5 million U.S. casualties. MacArthur's staff in Manila was saying it would only take 250,000. We eventually settled on a million."

"A million Americans?" I said.

"That isn't killed. The number includes those we expected to be wounded or maimed." "Only a million killed and wounded?" I echoed dumbly.

The Admiral smiled a thin smile. "The times were hard," he said. "We didn't look at things the way people do now. I'll tell you about the estimates process if you would care for another glass of wine."

I nodded, still stunned. That specific offensive would have ground up my Dad and all his buddies, and I might never have been born. More wine sounded swell.

YOU HAVE NO IDEA

• • •

CHIEF PETTY OFFICER GRAHAM JACKSON PLAYING "GOING HOME" AS FDR'S
BODY LEAVES THE WARM SPRINGS INSTITUTE FOR THE TRAIN STATION.

The Willow bar was bustling. There were some very attractive ladies at the bar, and some self-important young men attempting to chat them up. Sara the lovely Lebanese waitres was smiling her perfect smile under her delicately curved eyebrows, her raven hair shining. Andre the waiter circled solicitously. Peter managed the long bar with aplomb.

Everyone in the bar was thoroughly in the moment, just as thoroughly as Mac and I were in another, a humid place with still sultry air and shadows sixty-five years long, almost to the day.

I was warming to the topic. "I would like to hear some more about that," I said. "The whole estimates process. I mean, the decision to drop the Bomb on Japan was a result of Harry Truman making the business case about the cost-benefit in American lives, right?"

The Admiral pursed his lips and took a sip of his Virgin Mary. "I was just a junior officer, but I had unique access at the time and have done a lot of research since. I have a monograph about the end of the war that the CIA did. It is unclassified now, but it never published for public use. It is only about thirty page long, but the attachments are more than a hundred. I looked at it the other day, and read the minutes of the meeting at the White House that talked about options. Truman was there—he became President when FDR passed away in April. Not YN1 Harry Truman."

"I got that, Sir."

Mac smiled and counted the olives left in his glass. "Well, after Graham Jackson played "Nearer my God to Thee" on the accordion at Warm Springs and Harry S was sworn in, there was a lot for him to learn. He had never been in the loop for decisions like Vice Presidents are these days. In fact, he had only been in office for eighty days or so, and FDR didn't talk to him about squat."

"So, the Spooks come to him after he is sworn in and tell him about the Doomsday secret? Didn't that mean Stalin knew more about the Bomb than the President did?"

"That is what I understand, based on the subsequent revelations of the Soviet penetration of the Manhattan Project. We kept our heads down and prepared for Operation Olympic, the land invasion of Japan."

"But you mentioned the estimates process. Didn't that shape what Truman eventually decided to do?"

"Oh yes. We worked with the plans division of CINCPAC forward under Admiral Forrest Sherman. Admiral Nimitz had two hats to wear. For the Navy, he was CINCPAC. For the joint forces, he was Commander in Chief of the Pacific Ocean Area-CINCPOA. But of course, he had to deal with MacArthur, the Army Commander, whose HQ was in Manila. Sherman's guys had to go and coordinate with them as the air campaign ground on."

"So what did you do? What was your average day like?"

"We supported the estimates process and the planners. First, Okinawa had to fall. Most of that had been done before we got to Guam, and we couldn't bypass it."

"I remember the planning process. Sometimes the day of the big exercise would arrive and you realized something really important had to be done ninety days ago and you were totally screwed."

Mac laughed. "My roommate in the two-story Quonset hut on CINCPAC Hill was an army Captain named Hal Leathers. He did our ground estimates, and he thought there were 100,000 Jap troops waiting for us, and 2000 kamikaze aircraft ready to strike the Fleet.

According to the traffic we decrypted, the biggest battleship in the world, IJN *Yamato,* was getting ready for a one-way mission to beach itself on the island and use its 18-inch guns as static artillery."

"The Japanese were determined to make this so costly for us that we would seek options other than complete victory, right?"

"You have no idea. The civilians on Okinawa, like on Saipan, were indoctrinated to believe that the Americans would kill everyone on the island. Admiral Nimitz sent 1,500 ships, including some Brit fast carriers and a half million men."

I pursed my lips. "Let me get the timing straight. The invasion started in April, didn't it?"

"April Fools Day. We found Yamato on the sixth and sank her the next day. The Japs lost over 107,000 military and civilian on land and 4,000 sailors at sea. It cost us almost seven thousand soldiers and another five thousand sailors to the kamikazes. It was something entirely new in battle, and it was a real problem. The running battle went on almost to the 4th of July, but once we had a decent foothold we had a place for tactical aviation to stage from, and the skies belonged to us."

The Admiral reached in the pocket of his tan suit and pulled out a list. This is what we worked out with Iron Pants to hit the targets we thought were important.

I looked at the list, which seemed to be compiled from a Far East Air Force chronology. I studied some of the entries:

"July 10: 83 Very Heavy Bombers bomb oil facilities at Amagasaki.

July 13/14: 30 B-29's mine Shimonoseki Strait and waters at Fukuoka, ports at Seishin, Masan, and Reisui.

July15/16: 26 B-29's mine waters at Naoetsu, Niigata, Najin, Pusan, and Wonsan. 59 other B-29's bomb Nippon Oil Company at Kudamatsu.

July 15/16: 26 B-29's mine waters at Naoetsu, Niigata, Najin, Pusan, and Wonsan. 59 other B-29's bomb Nippon Oil Company at Kudamatsu."

The Admiral smiled. "Local Targeting. You have to break out our target list from the master activity chronology to understand what was happening. There was an awful lot of activity and our campaign gets lost in the static. The Air Force and the Joint Target Board prefer it that way, and the story of those brave aircrew that carried out the POL and mining campaigns."

"So, tell me: how did you work with the Air Corps planners?"

"We didn't, at least not directly. That wasn't our job. We provided target nominations and let them do their work. We were concentrating on estimates and supporting the planning process. That is why Captain Layton took us forward. The main event after Okinawa was Olympic, the assault on Kyushu, and we were battling the Army staff in Manila about what it was going to cost."

"You said Hal came up with 2.5 million American casualties for the invasion?"

Mac nodded grimly. "MacArthur's people said 250,000. There was a lot of back-and-forth and that is how we settled on a round one million. It put things off to November, but we were gearing up for it. There was no alternative except a negotiated peace that would have left the militarists in charge. No one knew about the Bomb except Admiral Nimitz, the Chief of Staff and maybe a couple others."

"They didn't even know the thing would work. What did they call it?" I searched my brain without any luck.

"The Gadget. No, they were pretty confident, but it wasn't proved until the day of the strikes at Kudamatsu. On the sixteenth of July, they blew up the gadget up at Trinity Flats in New Mexico."

"With the estimates of causalities I imagine there really wasn't much choice about it."

"Hal Leathers used to rave about the disparity in the numbers between us and MacArthur's staff in Manila. The Army actually started breaking some of the Imperial ground codes toward the end, and Hal was charting the units that were moving into Kyushu, which was the obvious next target. He had the real numbers of the real units, and that is why our assessment was so dire. There was one unit that he knew was present, but could not identify by number. He named it the "Leathers Unit." I have never seen that in the history books.

TRINITY TEST, ALAMAGORDO, NEW MEXICO. PHOTO: DOE

Hal was my best man after the war. He is dead now."

I tried to remember what they taught us about amphibious warfare. "Aren't you supposed to have a three-to-one superiority on an assault?" I asked.

"Four-to-one if you are General Montgomery," laughed Mac. "We could never get the numbers to work at more than 1.5-to-one. We would have been slaughtered."

He leaned forward. "We had a translator from FRUPAC named John some-thing-or-other. I'll think of his last name. He was slated to be in the first wave of the landings on Kyushu. He had a chance to actually visit the beach that his unit was going to hit. He looked at the caves and fortifications and realized that if Truman had not authorized the use of the Bomb, that is the very piece of sand where he would have died."

"Amazing what changed in just a few weeks," I said. "You don't know the half of it," chuckled Mac.

POTSDAM AND MONKFISH

• • •

PLANNING THE POST-WAR WORLD. POTSDAM LEADERS, JULY, 1945.

The dining room of Tracy O'Grady's Willow restaurant is a contrast to the bar area where I normally hang out. It is much lighter and airier than the dark wood area, though it shares a certain intimacy. It is all white tablecloths and spare elegant furnishings with solicitous staff. I do not make enough money to continue on from happy hour at the bar to the dining room as much as I would like. That is why Tracy established the $5 "neighborhood bar menu" to temp the regulars into a snack, even if we don't stay for dinner all the time.

This meal was something in the way of a special occasion, though. The Good Doctor was joining Mac and me for a meal after drinks.

He is always late, having *One Of Those Jobs* that involves talking to other *Very Busy People* whose schedules are frantic, and studied the menu, though it really doesn't matter at Tracy's Willow restaurant. Anything on it is good. If an item it is not composed of the freshest and tastiest of ingredients, it wouldn't be there.

I had the chateaubriand-for-two one night a few months ago, split with my older boy to commemorate some event, and it was better than anything I might have had at a specialty steakhouse. Creamed spinach and all the sides came with it, and the meal was extraordinary. I asked Tracy about it, since I haven't seen it since, and it was just something that came from *la boucherie* she frequents in the morning in her eternal quest for the finest ingredients.

The lamb, for example, is New Zealand and always good. I think it is one of the recipes she cooked for the Bocuse D'or competition in Europe, the one that required the custom serving set that now hangs in the narrow hallway that leads to the rear entrance and the restrooms past the private dining room.

I settled on the monkfish wrapped in bacon, which I would not have done anywhere else. The monkfish is supposed to be a real ugly creature in person, but much more approachable when filleted, sort of like life, or history.

Mac has a wonderful full-service dining facility at the Madison where he lives. He does not need the amenities and care that some of the other residents do, but he finds that not having to cook for himself is a nice convenience. Not that he can't mind you, as he would remind us.

When he comes to Willow, he drives the Jaguar from the garage under the building to a place he normally finds right in from of the restaurant, minimizing the risk of the crossing Fairfax Drive on foot.

This particular night I had strolled from the office and stared down the traffic from the dubious safety of the wide white lines that VDOT has lately begun placing with the apparent intent to confuse both motorists and pedestrians. Cars are supposed to stop for people, who have become emboldened and dart out unexpectedly in mid-block.

I made it almost all the way, edging uneasily in front of a gigantic and somewhat ambivalent Escalade SUV. Clear of the massive fender I looked up to see a hurtling bicyclist coming directly for me. My heart leaped into my mouth, and I jumped for the safety of the curb.

The city is plagued by these Lance Armstrong wannabees who brook no opposition to their speedy progress, and view the rest of us with contempt.

I suspect that is why Mac drives, even though The Madison is only across the street. Spry as he is, the years have taken a half-step off his best times.

He has had some times, and it is one of those extraordinary pleasures to have him as my own personal time machine. We had been talking about the estimates process that supported the planning for the invasion of Japan. I had a stack of bar napkins in the pocket of my suit jacket that I would try to unscramble later, and was increasingly aware that having lived the experience, Mac was one of the very few on the planet whose opinion was worth counting.

He decided to take the safe bet on the lamb. The Good Doctor opted for a table order of the Warm Gruyere Cheese Puffs with Black Truffle Sauce and one of the signature grilled flatbreads with calamari ali olio, garnished with roast garlic, three cheeses and oven dried tomato.

The Doctor is a historian by trade as well as a reserve Navy Captain, and he has been trying to unravel the history of the Air Intelligence trade. Mac is the last man standing who recalls what Admiral Forrest Sherman had in mind for the craft that emerged from World War Two, morphed through the chaos of Korea, and provided the front-line support to the Nuclear Navy of Admiral Arleigh Burke.

I was interested, too, since if there was something I would have put on my tombstones when this is all over, it would be: "Socotra. Air Intelligence Officer. Cold War, DESERT STORM, GWOT, OCO."

So, the Doc was trying to steer things around to Admiral Forrest Sherman, and I was determined to understand the last months of World War II.

When the waitress left us—a pert young woman in a white blouse and dark apron I did not recognize—I made a stab at keeping us in July of 1945. "I was just in Potsdam in May. It was raining like hell and my lovely associate and I looked like drowned rats by the time we hiked just a couple blocks from the train station. We had intended to tour Cecilienhof, the home of Crown Prince Wilhelm Hohenzollern where the Big Three held the conference to determine the future of the world. Never got there. We took refuge in the nearest church-St. Nikolaikirche. There was a room off the nave that had a brief history of the restoration of the place and I was blown away by what the town must have looked like when the wartime leaders arrived. It was mostly rubble."

Mac nodded. "I imagine so. It was a long way from where we were on Guam, and we knew that VE Day meant that VJ Day was inevitable. It was just that the Japanese didn't seem to agree."

Doc said that the militarists had been frantically been trying to negotiate for a cease-fire with the Russians before the Conference, and were willing to make concessions in Sakhalin, Manchuria and some other territory to get it. Harry Truman had been president for less than three months, and FDR, failing in health had been a softy as far as Uncle Joe Stalin was concerned.

"He once said he had a hunch that Stalin was not the kind of man to take advantage of him and that he would give him everything he possibly can and ask for nothing from him in return, noblesse oblige, and he won't try to annex anything, working together for a world of democracy and peace."

"That is a pack of crap as bad as President George W looking into Colonel Putin's eyes and seeing a good soul," I said.

Doc rolled his eyes. "There is a whole body of scholarship that holds that the U.S. cynically held the secret of the bomb in Truman's back pocket and sacrificed civilian lives at Hiroshima and Nagasaki to demonstrate to the Soviets that there was really a new world order and the U.S. was large and in charge."

"Those were the same idiots who wanted the Enola Gay exhibit on the fiftieth anniversary at the Air and Space Museum to be a big apology for American war crimes."

Doc nodded. "It was the curator of the collection, Jim Crouch, who said it could either be a 'feel good' exhibition, or a testament to the horror of war. It could not be both in his mind."

I said that everyone who saw it who had a dad who would have been there for the apocalypse would have disagreed with him.

"They started too soon," said Mac. "Too many of use were still alive who remember. They had to change the story-board to reflect the estimates of American casualties. "As you might imagine, I have followed this with a great deal in personal interest. In the CIA study of the end of the war are the minutes of a meeting at the White House with Truman, his chief of staff, Admiral Leahy, George Marshall, and Admiral King. They talked about all sorts of factors, but the one that was preeminent was the number of American kids who were going to die in the assault. Our estimates helped influence the course of the discussion."

"Did they talk about the Bomb? Did they all know?" I asked.

The Admiral looked at me sadly. "The President, Chief of Staff, Secretary of War, and CINCFLEET? Come on. Of course, they did."

"And the decision was to use the bomb."

"Of course. And it saved a lot of lives, American and Japanese, if our estimates from CINCPAC Hill were correct. I think they were. The revisionists will just have to wait until I am gone to say otherwise."

The waitress returned with another glass of that Chardonnay for me, one for the Doc and a glass of water, no bubbles, for the Admiral.

CAGEY FIVE

• • •

I was walking over to Willow when Old Jim called. I fished the phone out of my briefcase and answered—it was a District number, and I thought it might be him.

"Where are you?" he growled. "Mac is here."

"I am on my way," I said. "Getting ready to cross Fairfax Drive. Be there in a minute or two. Busy day. I will tell you when I get there."

Jim clicked off without comment. It had been a busy day. I flogged the Hubrismobile up to Shippensburg to confirm the burial space for Mom and Dad, visited Eby's Granite Works to order a headstone, toured a couple likely spots to hold the reception, and found a place that would block some rooms to accommodate the family that will attend the funeral.

I made it back with an hour to spare, and went through the office mail after the company system took twenty minutes to boot up.

Our crack IT staff has succeeded in their Information Assurance mission so thoroughly that the system is now almost impregnable to use by the employees. Brave New World, I thought, and hoping I could make the date with Mac on time.

People were drinking out on the patio. The day was that

nice, and I had the top down at the cemetery. It had been thirsty work, and I was gratified to see that Liz-with-an-S was back behind the bar.

"I worried about you," I said, slipping onto a stool next to Mac.

Liz-S gave me one of those radiant smiles as she topped up my glass. "Clean bill of health from the Docs," she said. "I was going stir crazy at the house."

"There is a lot of that going around," said Mac. "I drove fifty miles today getting my grand-daughter to some medical tests out in Fairfax. I am starved. Is there anything new on the menu?"

"There may be duck tacos," I ventured, "but they are usually out of them."

Mac studied the menu, and I got my pen and notebook positioned. Neither of us had eaten that day, so Mac asked for the duck, and I decided to go with the Pollyface Farms organic deviled eggs. "I brought you my recipe for no-fry eggplant parmesan," he said, sliding a wire-bound cookbook toward me. "It is something we did at the Arlington Hospital."

I read the title on the book: "Comforting Foods; Comforting Times."

"I could use some comfort. Wait—it is right here at the end of Liz-S's arm in this glass!"

Mac laughed. "I am feeling great. I walked over from The Madison."

"No kidding! You are getting spryer and spryer! We won't be able to keep up!" I looked up the bar.

Old Jim anchored the Amen Corner. John-with-an-H was

wearing a worn Carhartt Jacket rather than his usual suit. "I tele-commuted today," he said, looking at his Happy Hour red with satisfaction as Jasper topped him up.

"Me too," I picked up my pen after inhaling a deep draft of an impertinent white wine. "Now, where were we?"

Mac looked at me with a twinkle. "You never know where to start, do you? Why not at the beginning. That is a good place to start."

"Nah," I said. "I like to jump around. I think I might have ADD."

"You think?" growled Jim.

"The alcohol helps," I said. "With the word that the Navy is going to start using breathalyzers on sailors when they come on the ship. It is ridiculous."

"I heard that," said Mac. "Would not have been popular in my day." He took a sip of Bell's, a fine golden lager out of Kalamazoo, Michigan. His duck arrived on a rectangular white plate and he began to nibble on the dark morsels.

"It is like they want to make General Order Number One permanent," I said indignantly. "You guys won World War II and you had a wine mess at the forward headquarters on Guam."

"Yes, we did."

"Well, let me ask you this. Your Boss Eddie Layton almost punched out Admiral Richmond Kelly Freaking Turner onboard the USS *South Dakota* at the end of the war. Turner was drunk onboard ship."

"So was Eddie," said Mac. "There was probably more alcohol flowing on that ship the night of the victory as ever was poured on a man-o-war. People used to ask me how I tolerated him, but he was always OK with me."

"That is an interesting cultural snap-shot," I said. "What about Admiral Nimitz? Did he drink?"

"Never saw him do so," said Mac. "He may have had a glass of wine with dinner, but he certainly wasn't a booze-hound like some of them were."

"Did you ever socialize with the Fleet Admiral?" I asked.

"Not really. Well, wait, there was one time." He took a bite of duck and looked pensive.

"Was that on Guam?"

Mac nodded. "Yep. Admiral Sir Bruce Fraser—F-R-A-S-E-R" he spelled out as I scribbled, "He brought HMS King George V into Apia to present Nimitz with the Order of the Bath."

"Wow. Were you there for that?"

"Yes, I was. The Admiral sent out a note that anyone who had a set of dress whites could go out to King George V—Cagey Five, we called her—and attend the ceremony. I happened to have a set of whites, and I was included in the party. It was around August 12th, 1945, I think."

"Did the Admiral drink at lunch?"

"Dunno. He and Fraser went to the Flag In-port Cabin, and the rest of us were in the wardroom with the Brits. We had lunch, and then we drank all afternoon, and then through dinner."

"That must have been pretty fun. I have gotten smashed on Canadian warships, the last time being at Fleet Week in San Francisco in '98. I miss a civilized wardroom."

"This was on the way to not being so civilized, so we decided to get the boat and go back ashore. We were walking down the brow to board when we heard Cagey Five's 1MC crackle to life. The Captain announced that the Imperial Japanese government had made a decision to honor the terms of the Potsdam Declaration."

"What a moment," I said in wonder.

"Yes indeed. It may not have been official, but we turned around and climbed back up and had more drinks in the wardroom before we finally went ashore to sleep it off. On the way we stopped to ask Colonel Purple, the crusty old senior Marine on the staff if he had heard anything, and he said he hadn't."

"Wasn't that the guy whose house you flooded when your car hit the fireplug out front?"

Mac smiled. "Yes, it was. He didn't think much of the junior Navy officers."

"That was funny. But as to the merriment, you were entitled to it," I said. "It meant that everyone was going to live, and no one was going to have to die in the invasion of the Home Islands."

"Everything changed. Everyone had a different reaction, and most just wanted to go home as fast as possible."

"Except you."

Mac nodded. "I didn't have a job to go home to. I was single. I liked the Navy."

"It was a Navy that I remember, but is just a fading memory now."

"I think you will be surprised by that. Any institution that has survived a couple centuries will probably survive what is going on now."

"I hope so. I sure had fun in my Navy."

192 • • •

"So, the next day we resolved to host the Brits ashore in thanks for the open bar in their wardroom on Cagey Five. Then they had us back. It went on, back-and-forth, for four days."

"Sounds like fun," I said, taking a sip of happy hour white from a glass that never seemed to get dry."

"It was. But by the time the actual surrender was announced by Hirohito on the 15th everyone had been partying for days. I went up to the club at Nimitz Hill to have a cocktail."

"Must have been wild," I said.

Mac shook his head. "Nope. It was kinda funny. There was no one there. Too much merriment over the last four days, and no one came."

"Not quite what I would have expected," I said, putting two fingers across the side of my glass and winking at Liz-S.

"Well, that pretty much sums up the whole thing, in my experience," said Mac. He dipped the last bit of duck in a dash of hoisin sauce, and happily popped it in his mouth.

PUT UP YOUR DUKES

• • •

I was going to write about cars this morning—it is the time of the season when the people with whom I share the mad eccentric passion for the wheels of yesterday are out and rolling. But this date is also something significant on the life of the Republic and the world.

On this day in 1945, an official announcement of Japan's unconditional surrender to the Allies is made public to the Japanese people. It was not VJ Day, that is tomorrow, celebrated here on this side of the Date Line. But it is the day the Japanese war cabinet decided to throw in the towel, and I will let the vanquished express the anguish before turning my attention to the last man I will ever know who lived this.

The timeline is a little jerky: the War Council, urged by Emperor Hirohito, had submitted a formal declaration of surrender to the Allies on 10 August, though fighting continued between the Japanese and the Soviets in Manchuria, and between the Japanese and the United States in the South Pacific. Two USN ships, the Oak Hill (LSD-7) and Thomas F. Nickel (DE-587) were sunk by IJN submarine attacks east of Okinawa.

In the afternoon of August 14, Japanese radio announced that an Imperial Proclamation accepting the terms of unconditional surrender drawn up at the Potsdam Conference would be broadcast by the Emperor. The news was not met with universal acceptance. More than 1,000 Japanese soldiers stormed the Imperial Palace in an attempt to find the proclamation and prevent it from being given to the Allies. Soldiers still loyal to Emperor Hirohito held them off.

That was then. This morning, the Japanese continue to honor 14 August as the day their world changed forever. Prime Minister Shinzo Abe said that Japan's past "heartfelt apologies" for World War II will remain unshakeable in the future. He went on to express "profound grief" for all who perished in the war and acknowledged that Japan inflicted "immeasurable damage and suffering" on innocent people.

Abe's statement included these words: "On the 70th anniversary of the end of the war, I bow my head deeply before the souls of all those who perished both at home and abroad. I express my feelings of profound grief and my eternal, sincere condolences."

All that said, he noted that more than 80 percent of the country's population was born after the war and had nothing to do with the war. I don't know if he is serious about the notion that generations who are not responsible for the bloodbath be condemned to continually apologize to people who also were not there.

I will probably never again know someone who was there and lived the events that transformed the world, but I was lucky enough to have one of the best friends a guy can have. That is why I turn back this morning to the same week five years ago, and sitting at Willow with a Great American, one of a generation whose like we may not see again. We were with Mac, and the memories flood back with the breath of August in Washington, and the cool darkness of the Willow Bar:

The bill was paid, and people were starting to drift out of the bar area and over to the restaurant side of Willow. It was Restaurant Week, after all, and the *prie fixe* menu with choice of appetizer, entrée and dessert for $35 bucks is an attractive deal. Tracey O'Grady is always trying innovative things at Willow, and that is one of the reasons it is so much fun.

We need some fun. It could be that the recession is finally going to come to Arlington, which has ignored it thus far. At least the Defense sector has, what with two or more wars running through the decade. "Secretary Gates says he is going to retire next year, and take the national security succession issue out of the Presidential equation," I said.

Mac was straightening the stack of books and papers he had brought to the bar to aid the discussion. "He has done a fine job. One of our better Secretaries, in my judgment, and I have seen them all come and go, from James Forrestal on."

"It sure was a different world," I mused, looking at the empty wine glass in front of me. "I heard the Chinese are asserting sovereignty over the Senkaku Islands west of Okinawa."

"Oil, I would imagine," said Mac. "The islands were part of the U.S. military mandate that we gave back to the Japanese in 1972 along with Okinawa. The Administration doesn't seem to have the grit to explicitly tell the Chinese that they are Japanese territory and subject to the provisions of the mutual defense treaty."

"I guess we don't want to irritate the Chinese," I said. "But it seems like we are going to look like a Paper Tiger. If we cut carrier strike groups in the defense retrenchment, and the ASEAN nations are afraid to let us land-base the Air Force in Asia, what the hell do we think we have as a credible threat? Send a couple carriers against the whole of China?"

"Certainly is a policy conundrum," said Mac. "It is a complete circle from 1940, only with Japan playing the role of today's China. Article 5 of the 1960 "Treaty of Mutual Cooperation and Security" with Japan states clearly that it specifically applies to the Senkaku Islands. Try as Secretary Clinton might to soft pedal things, it is hard to get around that."

"It seems like we are surrendering the issue without discussion," I said. "Like when the Japanese strafed the American gunboat Panay on the Yangtze River after the Rape of Nanking in 1937."

"Yes indeed. And we did not respond, and four years later, the Fleet was on the bottom of Pearl Harbor. Take your stand, put up your dukes. It is pay me now or pay me later." Mac leaned forward. "I will give you a last story before we go."

I picked up my pen and flipped to a clean page on the notepad I had lifted as a souvenir from the U.S. Senate years before.

"Eddie Layton was one of the few guys who saw the beginning and the end of the war, said Mac quietly. "He started it in Admiral Husband Kimmel's office on the day of the attack. It is quite a professional tribute to his abilities that Admiral Nimitz took him along as his personal guest to the Surrender in Tokyo Bay. Eddie actually masqueraded as his bodyguard and carried a pistol. As a member of the Pacific Fleet Commander's official entourage, Eddie was quartered in USS *South Dakota*, (BB-57)."

"She was a battle wagon," Mac continued "New construction in 1939, and she had thirteen battle stars before it was all over. Seven hundred feet long and bristling with guns, she carried twenty-four hundred officers and men, plus the embarked staff, if there was one.

She had a Panamax beam on her, built to within inches of fitting through the locks on the Canal, so she was roomy and the wardroom was vast."

"I have been in the wardroom on Alabama, one of her sister ships," I said. "It was a lot different than the carriers I rode. Stately. I think they had cutlass racks on the quarter deck."

"They probably did. It was a continuation of an old tradition, carried almost to the ultimate level that the four Iowa-class battle-wagons represent. Halsey was embarked in *Missouri* in Tokyo Bay for the surrender, and Admiral Nimitz took South Dakota as his."

Mac chuckled. "She was a ship that set a unique record. A twelve-year-old managed to sign up for the Navy after Pearl Harbor, and he actually got assigned to her in time for the battles in The Slot. He was wounded, too. Calvin Graham was his name," Mac said with satisfaction. "You can look him up. He wound up getting a Big Chicken Dinner—a bad conduct discharge—when his Mother disclosed his real age."

South Dakota had the same sort of wild change of circumstance that everyone did. In early August, she was shelling northern Honshu and supporting carrier strikes against Tokyo right up to the last surface action of the War on the 15th of August. On September 2nd, she was anchored in the Sagami Wan with Admiral Nimitz and his guests on board."

"Eddie told me about it later. He was not senior enough to see the ceremony, but he said it was honor enough to be there. Following the signing of the instrument of surrender, the Allied delegates and spectators began to leave the *Missouri*. General MacArthur boarded a destroyer to be transported back to Yokohama. Admiral Nimitz left by boat after that and was soon back on board *South Dakota*.

Halsey stayed on his flagship and enjoyed a conversation with his old friend, VADM John

S. McCain, who would die of a heart attack just four days later. That would be Senator John McCain's grandfather."

"Stress," I said. "Pity to live just long enough to see the end of it, and not the beginning of the new world."

Mac nodded. "So here is where it gets interesting. Eddie Layton was playing acey-duecy in the wardroom of *South Dakota* when Admiral Richmond Kelly Turner strode into the room. "Terrible" Turner was in a state of high excitement. He was another of Admiral Nimitz' personal guests. He executed all the amphibious landings in the Pacific, from Guadalcanal to Iwo Jima, after all. But he was a son-of-a-bitch to his staff, and liked the bottle. Eddie said he was pretty fired up that evening. He started to shoot his mouth off, and the wardroom hushed at the remarkable sound of the inebriated four-star's booming voice."

"What was he saying?" I asked.

"He was going off on Admiral Husband Kimmel, of all things. Something to the effect that "Goddamned Kimmel had all the information and he didn't do anything about it. The court of inquiry said so, and they ought to hang him up higher than a kite!"

"But it was Turner himself who did not allow the critical Bomb Plot messages go to Pearl Harbor!" I exclaimed. "He must have known that. The court of inquiry was a white-wash to scapegoat Kimmel and General Short."

"Eddie sat there, stunned at what he was hearing. He had been there at the beginning, and was with Kimmel in the attack. Here at the end of it, the architect of the disaster was shouting that Kimmel ought to be hung up by his fingernails."

"I guess you can't do anything against four stars," I said thoughtfully, trying to imagine the scene with any of the other four-stars I have known in my life.

"Well, Eddie got pretty fired up, too. He corrected Turner in mid-rant. He told the Admiral that he had been there as Kimmel's intelligence officer, and he had been there in person on December 7th, 1941." "So what happened?"

"Eddie said the Admiral charged across the deck and grabbed him by the throat. Eddie was putting his dukes up to pummel the Admiral when the skipper of *South Dakota*, Emmet Forrestal, got in between them and broke it up."

I looked at Mac with amazement. The idea of decking a four-star Admiral made me admire Eddie Layton even more.

"Why on earth did Admiral Nimitz put up with Turner?" I said. "The man seems to have been unstable, and even if his war record was good, his poor judgment at the beginning is what contributed to bringing on the disaster."

Mac looked at me thoughtfully. "Admiral Nimitz always believed a man should get a second chance since he got one himself as a young officer after grounding his ship. Turner got his second chance after he screwed up in War Plans before the war. The cover-up of the whole ugly matter split a generation of senior officers and fueled a political and historical controversy over who was to blame for the Pearl Harbor disaster."

"Not to mention the fact that the war over who controls Signal Intelligence has continued right down to this year. We even had different officer designators for the Cryppies and the Intelligence officers, and never the twain did meet until sixty-five years later."

"I am not particularly surprised," said Mac with a deadpan face and a twinkle in his eye.

STAFF WORK

· · ·

I have a hard time keeping up with the Admiral. He is 91, and I am just a bit more than thirty years younger. He has the life force: I don't know what it is, but you can see it in his merry eyes.

I look in the mirror most mornings and see only blear in mine.

It was past all our bedtimes, but he had escorted all of us at the dinner table of the bustling Willow restaurant back to 1945, and being there with him it was hard to let it go. It can be a little disconcerting.

If you have read the 2003 novel "Time Traveler's Wife" you will understand the ability Mac has to transport you across space and time with his stories. I have to keep the notes on my cocktail napkin numbered, since earlier in the dinner we had visited 1955, jumping easily between the decades, and his creation of target folders for the A-1 SPAD drivers to study before launching against the Soviet Union with nuclear weapons.

It sounds preposterous now, but that was the case when the Admiral arrived at the FIRST Fleet and began to survey how training was being done to support the training mission for units going forward to the Western Pacific.

The Navy had fought hard to be included in the Single Integrated Operational Plan, the master scheme for the attack on the Evil Empire, should it have come to that. The SIOP (pronounced "sy-op") was an esoteric and highly political document that purported to de-conflict Air Force and Navy strike operations in the event that the balloon went up. It required a lot of Staff Work.

The Admiral was disconcerted to find that there were no materials to assist the dauntless men in their flying machines on their way to Armageddon. He fixed the problem in his tour by establishing a new staff to prepare highly sensitive target folders. There were no satellite pictures to help, as there were in my day, but at least the pilots had some way-points on the route to hell.

It is a magical thing, talking to someone else who was in the same very sensitive line of business a long, long time ago.

With Mac, it is as fresh as if it happened yesterday. The years fall away, and you can feel the presence of others, dead now, crowd around holding glasses of whiskey and nodding. The Admiral is their emissary, their guide between the worlds.

I could tell you where we were in the course of drinks and dinner, but mostly it was in 1945 since so much of our present rests on the foundation of what happened that year.

The Admiral recalls that the SPAD, the vaunted A-1 *Skyraider* that the Douglas Company built for the Navy (my Dad was a Spad driver before Mom made him get out of the reserves) was designed so that its internal bomb-bay could accommodate the dimensions of the atomic bomb.

I scratched my head at that. The Bomb was one of the biggest secrets in the world at the time, and certainly, it would not have been disclosed to the designers at their drawing tables at the Douglas Corporation. Or perhaps it was just a grim-faced staff officer in dress khakis who showed up one day after lunch, and spread his arms "just so," and told them it had to be that way, "just shut up and do it, you have no need to know why."

The Admiral was just a pup then, twenty-six and a Lieutenant on the staff, but filled with vinegar then as he is now. The war had moved west. Doomed USS *Indianapolis* (CA-35) delivered two very special packages to Tinian for assembly. Guam fell to the Americans in early August, 1944. Fleet Admiral Nimitz arrived to lead from the front and directed his staff relocate from Pearl to commence work there on the 15th of January.

Mac mentioned that the Marines were still catching eighty or more of the former enemy a day. They were hungry out there in the jungle, and sometimes the Marines killed them in the night, as the hungry soldiers scavenged for American food. The staff officers would walk by the bodies on the way to work in the morning on CINCPAC Hill.

They were planning the end game of the war, as best they could conceive it. The over-arching plan was called DOWNFALL, and included two major landings in the Home Islands. One would be led by General MacArthur in Kyshu, to the south, code-named

OYLMPIC, and a second one under the command of Fleet Admiral Nimitz on the Kanto Plain near Tokyo called CORONET.

"Why two invasions," you ask?

"One for the Army, and one for the Navy, Silly. They don't call it inter-service rivalry for nothing."

The Admiral was briefing events cribbed out of the Foreign Broadcast Intercept Service, which is called something a lot less ominous these days. That was really a cover, though, since his unclassified briefings were informed by highly secret decrypted intercepts of military and diplomatic communications.

If you are like me, history forms a jumble in my mental attic. For a lot of folks, amiable chowder-heads, it isn't even a jumble. It just doesn't exist. Here is what was happening that chaotic year of 1945, as the Admiral was briefing and planning.

Soldiers and Marines landed on Okinawa in March. President Roosevelt died on April 12th. The Nazis quit in May, and the troops were told to prepare for the invasion of Japan. The new fellow, Harry Truman, was informed that there was something being worked on, something big. Major combat operations were concluded on Okinawa in June, though scattered resistance continued.

The Scientists of the Manhattan District Engineering Project did not know if their bomb would really work, or if it would consume the atmosphere if it did. It was not tested until July 16th of 1945, as the Gadget was assembled at the hijacked McDonald Ranch and then trucked to the tower where the Los Alamos scientists predicted it would probably detonate with great force.

The CINCPAC Fleet Gunnery Officer, CAPT Tom Hill, was

sent to observe the event, and he brought a highly-classified film clip back to Guam to show Fleet Admiral Nimitz, for his eyes only.

Nimitz pursed his lips and kept his own counsel at the news as his staff planned the end. President Truman sent a question through his Joint Chiefs, once he knew what they had.

How many Americans were likely to die in the invasions?

It was a logical question, for a man who had options that others (except Uncle Joe Stalin) did not know about. In the Philippines and on Guam the planners paused in their deliberations and made calculations.

MacArthur's people in Manila low-balled the estimate. Maybe a quarter million, they said, ignoring the evidence of the communications intercepts that stated plainly that the Japanese knew where the landings would be and that everyone, man, woman, and child, would die to stop them.

The Admiral's team, headed by Ground Analyst Hal Leathers, looked at the evidence from the defense of Okinawa and calculated that it might take more than a couple million casualties to secure the capital.

MacArthur desperately wanted to command the invasion and damn the cost in treasure and lives. That was his way. It was like his insistence on receiving the Congressional Medal of Honor for good staff work so that he could join his Dad as the only father-son combination ever to win the highest military honor in the land.

There was discussion at the time back in Washington of considering Dugout Doug for promotion to a special "super rank" of General of the Armies so as to be granted operational authority over other five-star officers like Admiral Nimitz. Thank God the plans went forward as they did.

MacArthur would have been an American Caesar in reality then.

The story has been told of the days the world turned more acutely on its axis than normal. A-bombs fell on the 6th and 9th of August. I commend to you the account of the second strike on Nagasaki and the comedy of errors recounted by Major Sweeny in the Super Fort Bockscar, which resulted in an emergency landing on Iwo Jima to refuel and make it back to Tinian from the second most important mission the Air Force ever flew.

Maybe it didn't matter. Both of the weapons had worked as advertised, hundreds of thousands died, but not the tens of millions who would have perished if that astonishingly brutal path had not been chosen.

General MacArthur arrived at Atusgi Air Base on the 30th of August, the strangest month in the strangest year in human affairs to that date. I used to stay at Atsugi periodically and marveled at the revetments of old gray concrete that protected Japanese Ace Saburo Sakai's Zero fighters still ring the ends of the field.

They say there is much more still below in twelve great caverns carved out by the industrious Japanese, but it is too dangerous to go down there even after all these years, since traps were set with deadly efficiency, and now all those who set them are gone.

By the 2nd of September, the Allied Fleet was in Tokyo Bay to take the surrender.

Being so junior at the time, Mac stayed behind on Guam when Nimitz and four of his officers went to attend. His Boss Eddie felt bad about that, and so Mac was sent ashore to Yokosuka two days later, dispatched on an improvised and probably bogus mission to have a look-see at what they had accomplished.

Mac landed in a seaplane next to a tender moored in the Sagami-wan in the late afternoon, and a jeep took him to Yokohama where MacArthur's staff was preparing for the Occupation. For perhaps the only time in history, there was no traffic on Route 16 north from Yokosuka to Yokohama. There were crowds of Japanese on both sides of the road, looking impassively at the jeep impassively as it passed.

He arrived in the dark, and handed over his briefcase.

By the time he got back to Yoko, there was only time to trade a bottle of Three Feathers Whiskey from the wine-mess on Guam to a young Marine for one of three remaining battle flags 'liberated' by the Marines from the only Japanese ship that was still in the harbor, the battleship IJN *Nagato*.

Then it was onto a motor whaleboat to the seaplane tender for the flight in the morning.

"After we lifted out of the gray waters of the bay, the pilot did two long circles around the blasted capital before heading southeast for Guam. All the wooden buildings were ash, and only a few buildings stood in lonely isolation near the Imperial Palace. You could smell it."

"So, you got back to Guam and what was it like, Sir? Having it over and done with such abruptness? It must have been surreal. When did you go home?" "We were told we were to clean out our desks. We were flying on the Staff C-54 back to Pearl, direct, with a brief stop for fuel at Kwajalein Atoll."

"You must have accumulated enough points to be among the first to go home, back to CONUS, the Land of the Big PX," I said.

Mac dabbed his lips with one of the snowy white Willow cloth napkins. "Well, I did have a lot of points. More than most. But that is another story," he said.

I waved to the waitress, suddenly realizing I needed another napkin and either a brandy or a cup of coffee.

Or more likely, all three.

R-DAY

• • •

I took a sip of the rich Willow coffee, and countered it a moment later with a smoky taste of Brandy. The pile of napkins in front of me mounted. The Good doctor was having one of the extraordinary a *crème brulee* for which the restaurant was known, while Mac was having just the water.

"Even the idea of coffee, decaf or not at this hour is going to keep me up." I didn't doubt him; he always has more energy than I do.

"Samuel Eliot Morison wrote a postlude to his big history of the war," said the Doctor.

"The one that got so many things wrong about why things happened the way they did, since he didn't have a clearance. He said there were a remarkable number of postwar operations in the Pacific. Surrender of outlying Japanese garrisons, the occupation of course, minesweeping approaches to Japanese ports, and Operation MAGIC CARPET for the return of American armed forces to the United States."

"Yeah," I said, gesturing with my pen. "You mentioned you had a lot of points. How did that work, and why didn't you go home first? You had been in Hawaii since February of '42."

Mac smiled and took a sip of ice water. "One point per month of service; so I had over 50 points to start; I had been "overseas" for most of that, and each month counted for another point, and battle stars and decorations counted for more. Admiral Nimitz gave me a Bronze Star before he left command, and there is a citation for another one that I should have been awarded for the work we did to prepare for the battle at Midway. So, when you counted things up I was eligible to ride the Magic Carpet almost immediately. But of course we flew back to Pearl from the forward HQ at Guam."

"The first ones to go on from there were those soldiers and sailors who were on furlough in CONUS when the war ended," pronounced the Doctor. If they had more than 85 points, they were eligible for demobilization on the first R-Day."

Mac folded his hands on the white tablecloth. "The War Department had begun to plan for this job long before D-Day."

I swished some wine over my palette "I heard one time that the Navy went into the oval office to tell FDR that if the shipbuilding program continued as it was they would have 100 aircraft carriers in the Pacific, and the President told them to alter the plan."

"I imagine so, but we didn't know it was going to end.' "Well, someone must have."

"There was a fellow named Bill Tompkins, a Major General, who was in charge of the whole Magic Carpet thing. Apparently they waited until the day before the official announcement of R-day to inform the operational commanders what point total would make a servicemen eligible for release.

"They set it high at first, for the people in the European Theater, since they did not know how many troops would be required to conquer Japan. The Navy was specifically excepted from participating in the first flood of kids coming home after VE Day. The ships were needed to take the war to the enemy."

USS *MISSOURI* AND THE SURRENDER FLY-OVER. OFFICIAL U.S. NAVY PHOTO.

"Of course, with the surrender, the Navy jumped on things quickly. Magic Carpet featured every kind of ship in the Fleet. They welded bunks into everything, from aircraft carriers to battlewagons. New Mexico, Idaho, Mississippi, North Carolina, and two carriers plus a squadron of destroyers filled with thousands of servicemen from every branch of the military headed east.

The flotilla stopped at Okinawa to pick up more from the 10th Army. Aircraft carriers being the most popular with the soldiers because their size made them more comfortable rides than a destroyer or one of Henry Kaiser's Liberty Ships. The carriers were floating cities—they had had movies on the hangar decks, gyms, spacious lounges for officer and enlisted personnel, well-stocked ship's stores, fresh food, hot meals and three-to five-tiered bunks were installed on the hangar decks to accommodate the troops.

"I just did an interview with another vet. He was younger than you. Bill McCullough didn't have many points, since he just got to Saipan in November of '44. They shipped him up to North Field at Guam for a while before they let him go. He' was happy with traveling on an LST—a "Landing Ship Tank.""

"I imagine he was. I was back at Pearl then. Captain Layton had been called back to DC to testify in the Congressional investigation on who was responsible for the disaster at Pearl Harbor."

"That was the one where they pinned the blame on Admiral Kimmel and General Short?"

"Oh yeah." Mac scowled. "Never forget: Washington is always right. Joe Rochefort had been put in for the Distinguished Service Medal for figuring out the Japanese were going to attack Midway, the OPINTEL that allowed Admiral Nimitz to bushwack them and sink four carriers and a heavy cruiser. I broke the back of the Imperial Fleet only six months after the attack on Pearl. Washington thought the attack would be on Dutch Harbor in Alaska. Joe Rochefort was right, and the Redman Brothers were wrong."

"They really were bastards, weren't they?"

"Don't get me started. They managed to get Admiral Stark's Chief of Staff Joe Horne to not only deny Rochefort the DSM, but award one to Joe Redman. He got himself promoted to Rear Admiral, too, and his younger brother John to Captain, early. Admiral Horne was an ambitious son of a bitch. Admiral King never trusted him, even if he basically ran the Navy during the war."

"Didn't they force Rochefort out of Hawaii, too?"

"They forced him out of intelligence. They claimed Joe had missed signs of the attack on Pearl, and only Washington had everything right. By the time they were done, the most gifted code-breaker and Japanese linguist the Navy ever had finished the war as the commissioning commander of a floating drydock."

I shook my head in wonder at that. "Didn't you finally help set the record straight?"

Mac nodded. "President Reagan awarded Joe Rochefort the DSM in 1984. Unfortunately, it was posthumous. But Joe's kids were there to see it."

"I read Captain Layton's book. He really savages those assholes. But the damage was done, right?"

"Yes. Layton made admiral, but he dragged his feet at printing his story of the war. So much of it was highly classified. His book "And I was there" didn't come out until 1985, and that was posthumous, too."

"I still can't believe it was a drydock they forced him to. I imagine that ensured he would never get any recognition for the rest of the war. That was probably the only kind of ship that didn't help bring the troops home."

"True. It was quite remarkable. The Truman Administration could not stand the public and bipartisan Congressional outcry to bring the boys home. There were demonstrations by troops in Europe and the Philippines by the end of the year. Only a dust-up overseas could have stopped the flood going home, and it didn't happen. The Japanese efficiently disarmed themselves. December of '45 was the biggest month for the Magic Carpet."

The Doc furrowed his brow as he sat across the apex of the Amen Corner. "As I recall, nearly 700,000 servicemen came home that month alone. It was like someone had pulled the handle on a toilet to drain the tank. The American armed forces shrank from about twelve million in June of '45 to one-and-a-half million two years later. It would have a dramatic effect in the Pacific. By 1950, the North Koreans calculated they could push what was left of the American presence into the sea before we could respond."

"My Mom still talks about the boys from her hometown coming back through New York. The one thing she was not going to do was the same thing. She was not going to go back to the Ohio River Valley. The war set her free."

"It certainly did change things for a lot of people. But it was going to change again. Congress had passed a law that everyone who left a job to join the service would have that job waiting when they got home."

"So, with your high point total, why didn't you go home?"

Mac reached in his pocket for his wallet, I reached for mine and the Good Doctor reached for his, looking for all the world like a three-way Mexican stand-off.

"I told you everyone had a job to go back to, guaranteed by law. I had a bit of a problem. I had joined the Navy right out of the University of Iowa. I was one of the few people on the staff who had nothing to go home to."

I won't tell you who won the check, since you can probably guess. And I don't need to tell you that what happened next, after serving under three revolving-door Fleet Intelligence Officers, Mac found himself on a boondoggle back to Washington in 1946, the new center of the universe.

Remember, he was a deck officer who had never served on a ship, and the future looked limited. After the Wahoo was sunk with all hands, he had decided the submarines were not

for him. He was trying to figure out what the future might hold as he walked down the passageway of the third deck of old Main Navy by the Reflecting Pool on the Mall.

It was all luck again. He happened to see lean and energetic Admiral Forrest Sherman coming the other way, and the Admiral remembered Mac well from their days together on Guam, and greeted him warmly.

The Admiral put his arm around Mac, and said "I am concerned about all my Air Intelligence Officers. Most of the ones I recruited were lawyers, and with the war over, they have all gone back to civilian life. I need to start a cadre of intelligence professionals for this new world. We need to keep you, Mac. There is a selection board meeting right now to pick the first batch."

Mac looked at him, curious.

"Here is what I want you to do. Take thirty days leave and go back to Iowa. Then you come back here and contact my EA Captain Espey. He will tell you what to do."

Mac shook his hand. It was the last time he would see the Admiral in person, but it was just enough to determine the course of the rest of his life.

WILLOW HISTORY

• • •

It is sometimes said that the Department of War never lost one, and the Department of Defense never won a conflict. There is some room to quibble, since at the time we considered the first Gulf War to be a victory, since it liberated Kuwait from Saddam.

All I can say is that the last complete victory, ambiguous as it is, was fueled with whiskey and sustained by nicotine. I used to talk to Mac about that at the Willow Bar, though both he and the bar are gone now, just as most of the brands of that age are gone now; I have not lit up a Lucky in some years, and you cannot find Three Feathers whiskey anywhere.

Ships have the same problem. Regardless of how expensive or grand, they are nothing more than holes in the water that the owner, private or government, attempts to fill with money. Mac was a wonderful guy to talk to-his memory of the war years (and all the rest of it) was crystal clear, and he loved having a knowledgeable audience that could ask the right questions. Willow was the perfect place to hear about history.

Mac's team at Station HYPO tracked all the call-signs of the Japanese Fleet, and when Mac moved over to the Joint Intelligence

Center—Pacific Ocean Area (JIC-POA) he reported the general orders of battle in the weekly Fleet Intelligence Bulletin.

The Flagship of the Imperial Fleet, IJN *Nagato*, is a case in point. After Pearl Harbor, she saw action only once, during the battle of Leyte Gulf. She was a victim to a colossal error is strategy, which is the lifeblood of the successful war.

The Americans had adapted to the newest special weapon, the aircraft carrier, and only operated their battleships under the cover of aircraft. The Japanese never crystallized their doctrine in quite the same manner, failing in the integration of the combined-arms concept. The IJN leadership kept their major units in reserve for the decisive battle, which as it turned out, happened while they were waiting.

At the outbreak of the American phase of World War II, *Nagato* was under the command of Captain Yano Hideo, and with her sister Mutsu formed Battle Division 1. *Nagato* was the flagship of the Combined Fleet, flying the flag of Admiral Yamamoto. On 2 December 1941, *Nagato's* radio-room transmitted the signal "Niitaka yama nobore 1208," which translates from the Japanese as an admonition to "Climb Mount Niitaka on 12/08."

Mount Niitaka was the highest mountain in the Empire of the Sun, over 13,000 feet in height, and located on the island of Formosa, which we now call Taiwan.

That signal committed the Carrier Strike Force to the attack on Pearl Harbor, and Japan to expand its war in China to the rest of what they considered the Great East Asia Co—Prosperity Sphere. The glory was swift and fleeting. Admiral Yamamoto transferred his flag from *Nagato* in February of 1942, breaking it from the awesome new battleship Yamato's mast on the fifteenth.

Nagato's time as the most special weapon in the inventory had passed. Yamato and her class were stupendous, the first Japanese warships designed in the post-Treaty era. The limitations that preserved the primacy of her class had been established in Washington in 1922, and then extended in London in 1930. The international community hoped to put off another naval arms race until after 1937, at the earliest.

Yet another desperate diplomatic effort was scheduled for 1936, again in London. But under sanction, the Japanese withdrew from the pact. They commenced work on Yamato in 1934, and preliminary designs were accepted on March of 1935. After modifications, the design for a 68,000 ton ship mounting 18.1-Inch guns was accepted in March 1937.

Constructed under intense secrecy at the Kure Naval Dockyards, Yamato was commissioned two weeks after the attack on Pearl Harbor.

Nagato accompanied the First Fleet to the battle of Midway, but was consigned to the second rank. She saw no action at Midway, which turned the tide of the war, and returned carrying the survivors of the aircraft carrier Kaga to Japan.

As a part of the strategic reserve striking force, she proceeded to the anchorage at Truk in 1943, under the command of Captain Hayakawa Mikio. Upon the evacuation of that strategic base in February 1944, she steamed to Lingga, near the occupied British Crown Colony of Singapore.

Beginning in June 1944, she took part in the series of operations against the Allies that started in the Marianas, and continued in the Philippine Sea, and Leyte Gulf. *Nagato* passed through the San Bernardino Strait with other units of the Central Force, and engaged escort carriers and destroyers of the Task Group 77.4.3, the famous desperate fight of Task Group Taffy-3.

Under continuous air attack from the Americans, the Japanese fleet withdrew to the north, arriving at Yokosuka for refit and replenishment. The course of the war had changed. Lack of fuel and materials meant that *Nagato* could not be brought back into service.

In December 1944, the battleship was tied with heavy lines to the Koumi pier, next to the gigantic *Yamato* dry-dock where the American Carriers are now accommodated. Worse, she was "cold iron," with no boilers on the line. Lack of fuel meant that she could not get underway for evasive action. Accordingly, the ship was painted in an elaborate camouflage and her turrets were covered with wooden scaffolding as camouflage.

The New Year of 1945 was a dismal one. The Imperial Navy was shuffling personnel to match its remaining ships. *Nagato* was assigned as a coastal defense ship, isolated in Yokosuka, with the rest of the fleet in the Inland Sea.

In April, Rear Admiral Otsuka Miki arrived to assume command. Concealment operations continued, though increasingly it was apparent that battleships were hard things to hide. In late June, the word came that Okinawa had fallen, and the home islands were the last objective of the Americans.

Nagato's secondary and anti-aircraft guns were moved ashore to be placed strategically in the low green hills around the harbor. Tunnels were dug to provide shelter for the 1,000 crewmen who remained with the ship.

On the first of July, the fast carriers and supporting ships of Task Force 38 moved north from Leyte Gulf, and took up positions to launch aircraft against the Home Islands in July.

One of the prime targets of the combined American-British Fleet was *Nagato*. Admiral Bill Halsey viewed her as the very symbol of the perfidy of Pearl Harbor, and he was determined to sink her, in port if necessary, just as the *Arizona* had been sank at the foot of Battleship Row.

Nagato was tied up with her bow facing northwest at nearly right angles to the inner harbor entrance, starboard side to the pier. The huge yard crane I remember from my days on the waterfront is evident in the ancient reconnaissance pictures.

Inclement weather conditions forced cancellation of flight ops on the 17th, but the next day, the weather improved. The carriers Essex, Randolph, Yorktown, Shangri-la and Belleau Wood launched two hundred and fifty airplanes against Yokosuka.

The flight to the target was uneventful, with no resistance. That changed at Yokosuka, with fierce anti-craft fire erupting from the hills around the harbor, and from the ships themselves. The attack went on for around twenty minutes. *Nagato* took a 500LB bomb in the bridge, coming in at an angle from the port bow, slanting down.

The blast stopped a bridge clock at 1552, and shattered the pilothouse. Commanding Officer Otsuka, XO Higuchi and thirteen others were killed where they stood.

Months later, the compartment was still a mess.

My aeronautical engineer Uncle Jim was also a radio nut, among other passions, had brief custody (via the death of the American XO of *Nagato*, whose job was to steam the great battleship to Bikini Atoll for Operation Crossroads) of the very radio tubes *Nagato's* radiomen used to direct the attack on Pearl Harbor.

I wonder if the Wireless Museum people know that?

THREE FEATHERS AND A FLAG

• • •

RADM Donald "Mac" Showers is flanked by a pair of Socotras in the Hoyer Foyer at the Office of Naval Intelligence in Suitland, MD, on the occasion of Mac's donation of the battle flag of the IJN Battleship *Nagato* in 2012. The Foyer is so named irreverently in honor of the former House majority Leader Steny Hoyer, D-5th, for his efforts to have the new ONI building located in his district. Photo: USN

The proud battleship *Nagato* had been pinned down in port for all of the last year of the war, and would not get underway again before the end. She was pier-side at the Yokosuka Naval Armory, heavily concealed with camouflage nets and plywood structures in July of 1945.

Soon-to-be Fleet Admiral Halsey had a personal thing about the ship from which the orders to attack Pearl Harbor had been issued, and he directed the THIRD Fleet to conduct a series of air raids on the base in order to put her underwater. Task Force 38, under Vice Admiral John S. McCain, was the blunt instrument of choice, and his force included nine fleet carriers, six light carriers, their escorts and a thousand aircraft.

On 10 July TF 38 pilots struck airfields around Tokyo and claimed to have destroyed 340 Japanese aircraft on the ground and two in the air. No Japanese aircraft responded to this attack as they were being held in reserve to mount large-scale suicide attacks on the Allied fleet during the impending invasion of the Home Islands. Mac Showers was unperturbed by the revisionist history about the use of the atomic bomb in later years.

He and his pal Hal Leathers in the Estimates Section of the Forward Headquarters on Guam were reasonably confident that a million Americans would be killed or wounded in the assault, never mind the military and civilian population of Japan who stood in their path.

The strikes against Yokosuka were *Nagato's* last combat action, mostly serving as an American target.

Only two bombs actually hit the ship. One impacted the 01 deck aft of the mainmast, port side. It detonated at the base of the Number Three turret, and the occupying Americans of the Naval Technical Group marveled that although it had distorted the barbette, the turret was undamaged. The blast scar left a nearly perfect image of the rising sun flag on the surface of the armor plate. Worse, it had penetrated the ceiling of the lightly armored deck near the wardroom, killing over twenty men.

There is some lingering controversy about a possible third hit. Something, possibly a five-inch rocket, tore through the port side stern and passed through the Admiral's mess and out the other side without exploding.

The technical team reported there was a gouge on the surface of the dining table and markings that looked like they might have been made by teeth. The team was astonished that no one was killed there.

Between the two bombs, *Nagato* lost thirty-five officers and men.

The Allies lost fourteen planes and eighteen fliers, most of them over the harbor.

That was the last action of the war for the old battleship. Rear Admiral Ikeuchi Masamichi was recalled from retirement to assume command in late July due to the death of RADM Miki .

The newest Special Weapons of the Allies were employed against Hiroshima and Nagasaki on August 6th and 9th of August, respectively. At noon on the 15th, the Admiral called the crew to quarters to listen to the unprecedented broadcast of the Emperor, calling on his subjects to end hostilities.

On the 20th, Captain Sugino Shuichi, an active-duty officer, arrived on the ship, and there was a small change of command ceremony to make him the last Japanese commander. The position she occupied at the pier would be needed for other uses. Under his direction, *Nagato* was relocated to the Number One buoy in Yokosuka's inner harbor.

On the 28th, the Americans arrived. There are several accounts of who "captured" *Nagato*. The official legend is that "a boarding party composed of about 35 men from the USS *South Dakota*, symbolically on surrender day (Sep 2nd). The version some like to believe goes like this: "Many artifacts were brought back aboard USS *South Dakota*. The battle flag of the *Nagato* was acquired at this time."

Charles M. Cavell, QM1, USN, preserved a *Nagato* flag and donated it to the crew of USS *South Dakota*, and then the USS *South Dakota* Memorial in Sioux Falls, South Dakota.

Other credible reports of the disposition of the flag locker include this entry from the diary of the Skipper of the USS *Buchanan*, CDR Daniel E. Henry, USN, who reported this on August 30:

"Battleship *Nagato* boarded; San Diego docks at Yokosuka Naval Yard; first sighting of POWs; transferred 40 POW correspondents with horrifying reports on POW treatment. Awakened at 0700 and found the DD *Nicholas* (DD-484,) captained by D.C. Lyndon, my first classmate at the Naval Academy) was waiting to relieve us. They had met a Jap DD on the 27th and taken some Japs to Admiral Halsey on the Missouri. We proceeded to anchorage but found our berth occupied by transports busy sending Marines ashore at Yokosuka. We anchored and watched the show. American planes are landing at the field at Yokosuka, says the commodore, but I am not sure.

0830 our APD (USS *Horace A. Bass* (LPR-124) went alongside the BB *Nagato*, boarded, hauled down the Jap flag and hoisted ours. They report *Nagato* #10 boiler still warm, a diesel OK as is the anchor engine and steering gear. Other steam lines cut."

I think that may have been the first boarding right then. Others followed over the next few days.

We talked about the whole thing at the Willow Bar in Arlington where we gather to swap lies. There was more than one flag from the battleship, and our pal Mac got one of them. Now it is at ONI, and it hangs in a place of honor at out in suburban Maryland. ONI is a gleaming building of steel and glass, one of the last of the Cold War buildings to be erected to replace the crumbling old buildings that had served in World War Two.

THIS FLAG FROM *NAGATO* WAS BROUGHT HOME BY A USS *HORACE A. BASS* CREW MEMBER NAMED WILLIAM WILSON. IT MEASURES AN IMPRESSIVE 134 INCHES BY 69 INCHES AND WAS SOLD AT AUCTION BY MARK LAWSON ANTIQUES IN SARASOTA, NY, FOR $1,800 IN 2013. PHOTO: MARK LAWSON

In the course of its construction, a former Director of Naval Intelligence had the high-security facility designated an official depository for the combat art commissioned by the U.S. Government. Accordingly, there are some spectacular original paintings dotting the corridors between the anonymous cipher-locked officers.

One of *Nagato's* Rising Sun Naval ensigns is there as well, in a wooden frame, preserved behind glass.

It is a centerpiece that connects the young sailors and officers of ONI with their history. I asked Mac how the flag came to be in his possession as we sat at the Willow bar one lazy afternoon. He summarized it this way:

"By 1945, I was with Fleet Admiral Nimitz at the forward headquarters at Guam, doing estimates for the carnage that would come with the invasion of the Home Islands. I was slipping target nominations on the sly to General Curtis "Iron Pants" LeMay. Admiral Nimitz took his Intelligence Officer Eddie Layton—my boss—to the surrender ceremony on the *Missouri*. Eddie felt bad that none of the younger officers got a chance to see it after more than four years serving in the conflict. He cooked up a semi-valid requirement to send me with a courier package to Yokosuka just a few days after the surrender, and I jumped at the chance to go and see up close just what we had done."

"One of our shipmates was on a ship out there, and I decided to take him some whiskey, thinking that would be something useful. I went to the wine mess on Guam and got a bottle of Three Feathers Whiskey to tuck in my bag, and caught a flight on a seaplane to Yokosuka." "I have never heard of the brand," I said. "Did it go well with Lucky Strike Greens?"

"As a matter of fact, it did," smiled Mac. "So, I delivered the package of papers and walked around the base, marveling at the idea the conflict was really over. I could not find my friend, and didn't really want to take the whiskey back to Guam with me. That is when I saw *Nagato* out at the buoy in the harbor. She was the last of the big Japanese ships in port."

"My flag came from a Marine sentry on the Yokosuka docks (about 4 or 5 Sept. '45) who told me he was a member of the boarding party on *Nagato,* and that his duty station was the flag locker. With this convenience, he said he had "liberated" four Rising Sun flags, two large and two small. Desiring such a souvenir, I told him I only wanted one and would take a small one. Then we negotiated the price and the whiskey trade, and he disappeared for a few minutes."

"He returned with a brown paper package that looked about the right size. I tore open just a bit of the wrapping to confirm it was the Rising Sun design, we completed the deal, and I then caught my boat for the seaplane tender from which I would depart the next morning for my return to Guam. I didn't open the package until I was back in my private quarters on Guam, and then was pleased with what I had. I showed it only to Captain Eddie Layton, who confirmed I had a genuine souvenir."

"There's more that can be told, but that's the most authentic account of my procurement. In summary and in short, I'm sure there are other *Nagato* flags, but mine was clearly used, obviously had flown from the ship, and was from a believable source. More than that I cannot say." Mac raised his glass and took a sip.

"As I look back on the encounter and the bargain I struck, I now believe I could have talked the Marine out of all four of his flags in trade for the bottle of Three Feathers Whiskey I provided. But I was satisfied, I'm sure the Marine was, and I have no idea how he disposed of the other three flags."

Nagato was systematically plundered for the next few weeks, and formally stricken from the Navy List on the 15th of November 1945. Rust streaked the hull and the proud pagoda mast, and gulls of the

Sagami Wan rendered their opinion of the works of man in streaks of white down the superstructure.

But that was not the last of the story, nor even the best part. That was going to come when the Last Battleship got a new crew. An American crew, and in the process, the enlisted ship-fitters discovered the *Nagato's* store of grain alcohol, which may not have been Three Feathers, but suited them just fine on the last deployment of the battlewagon.

JICPOA FOOTNOTE

• • •

M ac was trying—with some success—to explain how quickly things had changed with the end of the war. I was stunned at the pace of the changes he described. One lieutenant was all that was left from the thousands who had toiled in Building 251/252. Wendy Furness was still around when I came into the business, and was a well-respected officer.

After the surrender, his job was to get rid of the captured Japanese military equipment that had been sent to the intelligence center for examination and assessment—what we later called "S&T"—or, Science and Technology.

High on the rim of Makalapa Crater sits Building 251/252, where the Joint Intelligence Center Pacific Ocean Areas (JICPOA) collected, evaluated, and disseminated strategic and tactical intelligence for the Commander in Chief, Pacific Ocean Areas throughout World War II.

Five decades later, the building was in need of more than a simple renovation. Part of its foundation, built on fill, was gradually sinking into the Crater when Mason Architects, working with general contractor Nan, Inc., was commissioned to design its renovation and oversee its reconstruction.

Compaction grouting, according to the brochure, was used to stabilize the soil beneath the building, along with new concrete and masonry foundation walls. New plywood shearwalls were then installed to bring the structure up to current seismic code, and a new wood skirt was nailed along the perimeter to match the original.

Post-World-War-II additions to the building were removed and the interior was gutted and rebuilt.

Blocked-out windows were reopened, windows added over time were removed, and new fixed-glass windows sized to match the originals were placed in the original window apertures.

Nearby, at the intersection just beyond the Pacific Fleet Headquarters building, at the intersection of Halawa Drive, there is a fireplug surrounded by an imposing concrete circle.

I recall it from my days at FOSIC PAC; the hydrant sits slightly above the level of the senior officer's quarters behind it. In the war years, Colonel Purple, the senior Marine on FADM Chester Nimitz' staff, occupied the house.

One night, just after dark, a young officer from FRUPAC was headed back to his quarters in Little Makalapa. In the days of the black-out, there were three lines painted on the roads to follow, since headlights were prohibited. The custom was to follow the line in the middle. Fatigued from his labors, the young officer was following the right-hand line, not the one in the middle.

He collided with the fireplug. He was able to back his car off the offending infrastructure and proceeded back to his quarters.

Regrettably, the weakened plug later failed and flooded out Colonel Purple's residence.

Admiral Nimitz asked about the incident in the Command Briefing the next day, and Eddie Layton called over to Jasper Holmes to ask if anyone had reported the problem.

Mac sheepishly acknowledged the incident. Despite the irate Colonel, the war, and Mac's career, turned out just fine.

PART TWO:
HOT & COLD RUNNING WARS
WASHINGTON, LONDON,
NAPLES AND THE FORT

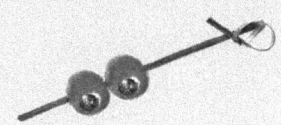

TIGHTENING THE BELT

• • •

SWISS COTTAGE TUBE STOP, JUBILEE LINE, LONDON.
PHOTO: LONDON TRANSIT SYSTEM

I f I were not in the act of taking notes, this would have been an ordinary late afternoon on a workday, seated at the bar at Willow, drinking. Instead, we were continuing the oral history interview with Mac, which lent the enterprise a certain high-minded quality. Liz—with-an-S, the lovely afternoon slack-time bartender approved and was solicitous. When the shift changed, and Peter and Jim arrived to service the industrial strength crowd, we had ventured a little off track.

"So," I said, taking a sip of my refreshing white wine, "The draft proposal of the President's bipartisan commission on reducing the federal debt calls for deep cuts in domestic and military spending, starting in 2012. That is going to screw the business climate for contractors.

"I presume the assumption is that the current economic crisis and the wars will be over and enable us to do even more irresponsible things in 2011."

"Like extend the Bush tax cuts, which would let me try to keep paying down the hit I took on real estate in the bubble. I have never missed a payment but I can't qualify to refinance because the loan-to-value ratio has

"Reform like the report recommends won't happen," said Mac, raising his amber glass of Bell's. "It is a draft. A trial balloon to see what gets people all riled up. We have been talking about fixing the tax code for a generation and it just gets more convoluted. From what I read, the changes would erase nearly $4 trillion from projected deficits through 2020. I will be 101 that year."

"I heard the plan is going to reduce Social Security benefits to all of us Boomers, though the people that don't pay taxes now—the poor ones, not the rich ones—will get higher benefits. They also propose eliminating the cap on the FICA deductions from our paychecks. Bastards. That means whatever the benefits I get back are just about what they take away to begin with."

"It is all in the art of the budget. They are going to promise you that with some current sacrifice, Social Security will be solvent for at least the next 75 years."

"They never seem to do the sacrifice," I said. "It is always us.

You will be 176 then," I said, making a calculation on the napkin. "You have been drawing Social Security for 26 years."

"Pays to have your timing right," said Mac with satisfaction. "But you get what you pay for."

The notes, as I review them this morning, begin in an organized enough fashion. We plowed a little familiar ground to get calibrated. I was prepared to listen to 1953, an settled into the rhythm of the long ago.

Mac wed Billie in 1948, an excellent year for new beginnings, and elected to stay in the Navy. After a period at the Pentagon writing the Political Cable, he was transferred to the European Navy headquarters, CINCNELM, which had taken over Ike's old SHAEF headquarters on North Audley Street near the Embassy in London.

THE SHOWERS' RESIDENCE ON MARESFIELD COURT. THE THIRD-FLOOR FLAT MUST HAVE BEEN TINY-MAC AND BILLIE LIKED THE ONE ON THE FIRST FLOOR.

I had intended to discuss the great tides in international affairs that were going on then; the Greek Crisis, the impact of the Marshall plan, the strange events in Iran.

Instead we got off on beds, though not literally, of course.

Mac and Billie did not take much to England except their beds and the 1948 Mercury sedan. It had been a bad war for the British, and they accepted a lot of things we wouldn't as a matter of necessity. Rationing was still in effect, though of course the Yanks managed to have their own supplies.

For one thing, the British slept on appalling mattresses, or at least that was the word in Washington before they decamped for London, and the beds were shipped to a row house located at 18 Maresfield Gardens, near the Swiss Cottage Tube stop.

"The station was opened just as the war was starting in late 1939 on a new section of deep-level tunnels constructed between Baker Street and Finchley Road. They used them as shelters during the Blitz. It is named for a nearby pub, an old one that dates from before the battle of Waterloo. It was originally called The Swiss Tavern, and later renamed Swiss Cottage. The pub is still there, or at least it was in 1950."

I made a note to check and did this morning. It is. "What sort of place did you live in," I asked, spilling a little wine on my note napkin, making the ink bleed.

"It was a nice place. It had been an imposing three-story home that had been divided into three flats. We had the ground floor, with the garden, and with the good American mattresses beneath us, life was pretty good."

"What was interesting was that Sigmund Freud, the pioneer psychotherapist, had lived just two doors down the row. He was dead, of course, the cancer having taken him in September of 1939, just a couple of months before the tube station opened. He had to get out of Austria with the growing madness there and lived the last year of his life in the relative safety of London."

ANA FREUD, MAC AND BILLIE'S NEIGHBOR IN SWISS COTTAGE.
PHOTO: BETTMAN ARCHIVES

"The house survived the Blitz, and his daughter Ana lived there most of the rest of her life. The house is a museum now, but in my time, it was just Ana's residence, where she carried out her life's work refining the principles of her father. "

"Freud didn't exactly invent the idea of the conscious versus unconscious mind," I said, working my slow way into unconsciousness, "but he certainly was responsible for making it popular. He also famously asked the question: "What do women want?"

"He never got a satisfactory answer," said Mac. "But Ana had some ideas about it, and so did Billie."

In a roundabout fashion, we had got onto the topic of bagpipes, which meant one of my napkins might have gone missing. We were moving backward in time, or maybe the notes were out of order. The pipes and piping were funded in Mac's home state of Iowa through the Department of War. The mechanics of playing the bagpipes involved both the Conscious and Unconscious minds, which had to be trained to work in concert.

Mac said you have to squeeze the bag with your left arm while blowing into the pipe and fingering the chanter. I thought that it sounded a lot like trying to strangle a cat while getting it to purr, but that in turn, was tied to life on a farm in Iowa, in the Great Depression.

We had jumped right across the war and into a time when the banks were closed, and there was no money at all.

"Imagine an America without cash!" said Mac. "That is a "Black Swan," even for you! Everything was done in barter, vegetables and meat for dental services, professionals and farmers alike getting by as best they could."

That was where the pipes came in since Mac played for three years in the Drum and Bugle Corps at the University of Iowa before he had to drop it to assume the duties of City Editor at the Daily Iowan, the campus newspaper. He might have been a reporter if the world had not lost its senses to the Monsters.

There was another glass of wine, and a Bell's IPA in there somewhere, though my notes are not clear as to the timing.

What is clear is Mac's recollection of change. His first president was Herbert Hoover, who inherited the first three years of the national disaster after the stock market crash in 1929. Those of us that did not live through it tend to see the thirties in scratchy black-and-white, flickering images of marching Germans and indomitable

Franklin Roosevelt propped at a podium, easing the crisis with bold new programs like the Civilian Conservation Corps and the Works Progress Administration.

"I was an FDR supporter," he said. "But I still went down to the depot to watch Mr. Hoover's campaign train come through, and listen to a speech the president gave in his doomed campaign for reelection in 1932. He didn't have a prayer, any more than this bipartisan panel's recommendations do. We could save ourselves, but I suspect we won't."

"That is the way I feel," I said, drawing a line under the words "mortgage interest deduction.

"Hoover was trying his best to turn things around, and he might have succeeded since all these great economic things are cyclical, and largely beyond the power of any one President to alter. Mr. Roosevelt was selling hope, and change, and that is exactly what people wanted. That is how Obama got elected and why the Republicans took back the House."

"It is all about the public's conscious mind," I said. "But Dr. Freud could tell you a lot about all the seething Id down below. "

Eventually we got to the point where my notes no longer make any particular sense. The bar at Willow was filling up with vibrant pre-weekend noise, and it was time for us to move on.

I walked Mac out to his gold Jaguar, and we made arrangements to meet again next week.

I walked across the street to the office tower where I had to retrieve the '04 P71 Crown Vic Bluesmobile from the garage under my office and re-enter the world of now.

The prospects were not encouraging. The cleaning crew, all Hispanic, was in the office when I ducked in to grab my briefcase and check the last e-mail of the day. It didn't make any sense, and I shut down the computer. 2012 looks like it is going to have some challenges. I wondered if I should go home, or just go back to the Willow and get ready.

I think you know that I did the right thing.

SLIVOVITZ

• • •

I had lunch with Annie, the den mother of our fractious group of brash officers who inhabited the back room at the Navy Office of Legislative Affairs in the Pentagon. There were some stories to be had in that office, let me tell you. They were about the trips we arranged for Congressional members and their staffs, and tales about all of us, and what astonishing things we had done representing the Department of the Navy on behalf of you, the taxpayer. We talked about some of that, but we talked mostly about Annie's Mom, a marvelous lady to whom I had been introduced a decade or so ago. I had a chance to meet a decade ago. Like Mac, she is gone now, and I am glad I had a chance to meet her when I did and hear some of her story. See, she was living in Belgrade, capital of Jug-Land, when Mac and his party were dining with Tito over at the White

Palace. She was the only other person I will ever meet that has that distinction, except of course Annie herself, who was sitting across from me at the Two Chefs Restaurant on Lee Highway. It was a great lunch. Annie has some stories herself, and who knows, maybe she will share them sometime.

Did you ever hear of Slivovitz?

It is a plum brandy that is made in Eastern Europe, and notably Yugoslavia. Rene LaPlante, Annie's Dad and Assistant U.S. Army Attaché could have told you about it, since it is one of the tools of the trade. "Plum brandy" sounds like something your great aunt might break out with the holiday fruitcake, and she might, if she was a heavy-duty woman.

What she would be bringing out was rocket fuel. Distilled properly, Slivovitz is more than a hundred proof and can have the impact of pure grain alcohol. It has loosened more tongues than anything this side of fiery Italian grappa, which is how it relates to Evelyn and Rene and the story I just heard from my pal Mac, who is a long-retired Spook.

Slivovitz and Tito is where the story goes, but it is very strange and needs a bit of context to really appreciate.

Evelyn LaPlante got out of the spy business in 1947, just ahead of the "*Informbiro.*" That is the term the Jugs use for the breakdown in relations between Stalin and Josef Broz Tito. The word is shorthand for the Communist Information Bureau, the Soviet-dominated propaganda organization that set the policy line for the new Soviet satellite states of Eastern Europe.

Some of the clients had actually wanted to attend the Paris

Conference on the Marshall Aid Plan, and Stalin obviously could not countenance that.

The Bureau had a brief run in Belgrade, but by 1948, it packed up and left. Tito was officially denounced by Moscow, and accused of "departing from Marxism-Leninism" and "exhibiting an anti-Soviet attitude."

These were serious accusations, though they sound a bit quaint today. If you add a *lot* of Slivovitz they drip with menace. That was reflected in the situation that convinced Evelyn that Stateside was the place for her and her little daughter Ann. Tito would open and close the border capriciously, in tune with a rhythm of the dance with Moscow. She got out when the opportunity presented itself in 1947, the year the Information Bureau was established and things were very edgy.

Stalin contemplated sending assassins to take out Tito. It was said when the charismatic leader in Belgrade was informed of the plot, he smiled grimly. He told his advisors that if anyone came for him, there would be two Jugs headed tight back for Moscow, and that would be the end of *that*.

There was a lot of bravado contained in that threat, and Tito cast about for the pivot point in the crisis with the Russians. Ann's Dad, Warrant Officer Rene LaPlante stayed in Belgrade as the crisis deepened, and was sent home when the Informbrio was expelled, and party purges of alleged Titoists throughout Europe began. Rene's services were needed elsewhere.

It took a lot of Slivovitz to consolidate good order in the new

Soviet empire, and nearly three years for Tito to culminate a bold strategy to balance the power of the East with the Power of the West. That is how my pal Mac entered The Balkan Problem came in 1951.

At the time, Mac was on the staff of the Commander in Chief, U.S. Naval Element-Europe. The command was normally located in London, in Ike's former headquarters on North Audley Street. At this moment, though, it had re-located to Naples, where he and his family lived in an elegant if somewhat threadbare apartment on the Napolitian economy.

The Jugs approached the Americans in Belgrade, and indicated that Tito wanted to talk.

Something dramatic was cooked up in short order. In December 1951, Mac was a Lieutenant Commander, and he was nominated for the special collection mission by the SIXTH Fleet intelligence officer, CDR Fred Welden. With LT Art Newell, Mac reported to the fleet flagship, the heavy cruiser USS *Des Moines*.

USS *Des Moines* (CA-134) in the Adriatic, 1951. Photo: USN

Vice Admiral Matthias Bennett Gardner flew his Fleet Commander's flag on the 17,000-ton cruiser. She was a remarkable floating statement. Roomy enough to handle the requirements of the embarked staff, she cut a suitably martial profile. She bristled with a main battery of nine improved eight-inch guns, twelve five-inch turrets on the superstructure forty-four assorted anti-aircraft guns.

With a bone in her teeth, she could make thirty-three knots and keep up with fast carrier and carried a crew of 1,800 officers and Sailors.

She was just the platform for this special assignment. She was about to make the first American naval visit to Yugoslav waters since the end of World War II. It was a ceremonial visit, which Tito intended to stick right in Uncle Joe Stalin's teeth. It was a bold move for the Jug leader, taunting his former mentor in Moscow, and the U.S. Navy was happy to accommodate him.

Just before Christmas, the Cruiser got underway and made for the ancient Croatian port of Rijeka, sometimes known by its Italian name of Fiume.

There was going to be a mission of extraordinary daring, and it would feature Slivovitz before dinner, slivovitz during dinner, and Slivovitz toasts of friendship in the drawing room after.

That is the story that became a chapter in Mac's biography, how Mac came to have dinner with Marshall Tito, the leader of all the Jugs, and the extraordinary party at the residence of the Naval Attaché the last night Des Moines was in port.

The power of Slivovitz cannot be underestimated. Diplomacy used to be a lot more muscular, and certainly a lot more fun, you know?

THE REVOLT OF THE ADMIRALS

• • •

CONVAIR B-36 PEACEMAKER, 1949. OFFICIAL AIR FORCE PHOTO.

"The Bomb got us all home," said Mac. "It saved us the hundreds of thousands of casualties, American and Japanese, that would have come if we had executed Operation OLYMPIC. That is certain. But once the bomb was used, the genie was out of the bottle." Mac looked down at his ginger ale a little pensively.

"James Forrestal began to agitate about the threat from the Soviets while the war was still raging. He visited Ike in England during the war to emphasize his concerns about the Soviets. Apparently, Ike agreed with him, though he had to walk the line in the Alliance.

The Soviets had sacrificed so much on the East Front, it was difficult not to accept that they had a moral justification for what they were doing to Germany."

"And it was better for German and Russian kids to die for their countries than ours. Didn't Patton say something like that?"

Mac grimaced at my paraphrase of the quote from Old Blood and Guts. "Victory is victory. But Forrestal certainly was proven right by the Communist political campaigns against the elected governments of Greece, Italy, and France after the war. The Truman Administration did not agree with Forrestal's concerns, though that was what drove Naval Intelligence to begin to track what the Soviets were up to. It was still the deepest secret in the OPINTEL business that our former Allies were the most important target. It was at the heart of everything in the struggle about unification of the services."

"So what was it like?" I asked. "Was the controversy as big as when Secretary Gates suggested eliminating the Marine Corps?" I could feel the effect of the happy-hour-priced *pinot grigio* and my scrawl in the notebook was getting wilder and wilder. Peter offered to top me off, and I put my hand over the top of the glass. This was important and I wanted to get it straight. I wanted a cigarette, too, but those days are long gone in Virginia's bars and restaurants, as gone as the 1940s themselves.

"You have no idea. Now, remember I was just a fresh-caught Lieutenant Commander then, but there were some amazing shenanigans here in Washington. It wound up in 1949 with the firing of the SECNAV and the CNO."

"Wow. You said it was about the budget more than anything."

"Of course. Everything here is driven by money. Anyone who tells you it is not the green-eyeshade crowd that drives strategy is a fool. The "Revolt" had been building for several years, but it climaxed the year we moved Y-Branch to the Armed Forces Security Agency at Arlington Hall Station.

CHIEF OF NAVAL OPERATIONS LOUIS E. DENFELD

SECRETARY OF THE NAVY JOHN L. SULLIVAN

"The drive for unification began in 1943, and the National Security Act of 1947 made it formal. The generals of the new Air

Force announced that strategic bombing, particularly with nuclear weapons, was the sole decisive element necessary to win any future war; and was therefore the sole means necessary to deter an adversary from launching a Pearl Harbor-like surprise attack or war against the United States."

"That is a breathtaking assertion," I said. "Like some of the things Secretary Rumsfeld used to say."

"Well, it certainly was a transformational doctrine. The Air Force leadership proposed that it should be funded by Congress to build a large fleet of long-range strategic heavy bombers, beginning with the B-36 Peacemaker bomber."

USS *UNITED STATES*, PICTURED IN DRYDOCK WITH HER KEEL LAID.
OFFICIAL U.S. NAVY PHOTO.

"I have seen the one they have at Wright-Patt Air Base. It is impressive."

"You can imagine that SECNAV Forrestal was opposed to that, and he continued to oppose it when he became the first SECDEF.

The Navy's position was that the triumph of the aircraft carriers in the Pacific War demonstrated the importance of fielding a new class of super-carriers, the first of which was to be the USS *United States.*"

"So the fight was between weapons systems? I said.

"Only so much money to go around," said Mac. "Something had to give, and when Truman appointed Louis Johnson to replace Forrestal, the balance shifted to the Air Force. The Navy leadership believed that wars could not be won by strategic bombing alone, and that preemptive use of the A-bomb was immoral."

"Though the carrier would have carried atomic weapons, of course."

"Of course," said Mac. "Less than a year after he took office, and without consultation with the Congress, he cancelled the *United States.* Then he announced that all the aircraft in the Marine Corps would be transferred to the Air Force, just as the Army's had."

"Good golly. That must have caused an uproar in Congress."

"It did, and several admirals resigned in protest. See, the Air Force could not control Naval Aviation, and Johnson sought to shut down procurement of new platforms not controlled by the blue-suiters."

"Obviously the Navy couldn't take that laying down," I said indignantly.

Mac clasped his hands. "They didn't. A little group called OP-23 was stood up by 31-knot Burke, the destroyer hero from the Southwest Pacific. Arleigh was just a Captain then, and the group began to quietly collect operational intelligence on the flight characteristics of the B-36. What do you know, an anonymous document then hit the offices on The Hill, claiming that the B-36 was a "billion-dollar blunder" and alleging fraud on the part of B-36 contractors."

"Nothing has changed, has it?" I said, smiling. "Do you think the Good Doctor will ever finish the biography of Admiral Burke?"

"Maybe. But this is part of that story, too, and gets better," said Mac. "The document accused Secretary Johnson of having a direct conflict of interest, since he had been on the Board at Convair, the airplane manufacturer that built the Peacemaker."

"That is hardball," I said. "Amazing."

"RADM Dan Gallery, the guy who captured the German U-Boat 505 and the Enigma coding machines, went on the offensive, too. He published an series of articles in The Saturday Evening Post. The last one was called "Don't Let Them Scuttle the Navy!" It was so inflammatory that Johnson wanted Gallery court-martialed for gross insubordination.

CNO Denfeld got caught in the crossfire and was dismissed, and SECNAV Sullivan was asked to resign."

"I wondered why Gallery never got beyond two stars," I said. "I never heard that he had been tried in court."

"He wasn't, but the articles ended his career, even if he was one of the more noted heroes of the surface Navy."

"So what happened to resolve the revolt?"

"A circus in Congress in 1949. The House Armed Services Committee found no substance to the charges against Johnson, that the anonymous paper was erroneous about the B-36, and the Navy civilian who penned it should be fired. The Committee held that the Army and Air Force were not qualified to pass judgment on aircraft carrier design. Arleigh Burke and Op-23 were not fingered, so the greatest CNO in Navy history kept his career, and eventually transformed the service into a modern force."

"And this probably suggested to someone that the Joint Chiefs be strengthened to keep the children from fighting."

"Omar Bradley, the last five-star of the nine appointed, was Chairman. He called the Navy a bunch of "Fancy Dans.""

"I imagine the whole argument shifted when the Soviets detonated their first bomb in December."

"And it completely changed the next year when the North Koreans came south. But Service unification really didn't really come to pass until Goldwater-Nichols was passed thirty five years later," said Mac. "But at the time, I was more concerned with shipping my car to England to worry too much about the next class of Aircraft carrier, but of course, you know which one she was, and for whom she was named."

I nodded. "CV-59, the USS *Forrestal*. My home from 1988 to 1990."

Peter put the black folder with the tab on the table in front of us. Mac reached for his wallet, but I grabbed the check. "You had three ginger ales, and I got all the wine," I said. "I was up in Philly a few weeks ago and was surprised to see that *JFK* and *Forrestal* are up there at Pier 4 at the Naval Shipyard. I hear FID is going to be sunk as a fishing reef."

"Easy come, easy go," said Mac, as we rose to leave.

"I still want to hear about your meeting with Eisenhower." "Then I imagine we will have to come back, won't we?"

USS *Kennedy* (CV-67) left, and USS *Forrestal* (CV-59) right, being prepared for scrapping. Photo: U.S. Navy

LIKING IKE

• • •

OFFICE IN PARIS, 1951. PHOTO: U.S. ARMY

M onday is a fine day to stop at Willow after work in early October, and I had a lot of questions for Mac. He enjoys the $5 bar menu and a ginger ale or two, and he is hungrier now that he has engaged a personal fitness trainer. He is determined to get out and about more, now that his arm has healed, and though he drives to the restaurant from the Madison Assisted Living high-rise across the street, he is looking fit and feisty these days.

Mac favors an aloha shirt under his sport coat, sort of Hawaii business casual, and he was seated at the corner of the bar next to

Old Jim, who is probably twenty-five years younger than Mac. Big Jim the burly bartender was talking about what the Steelers did yesterday.

They won, 28-10, over the Browns, and for the record, he was feeling pretty expansive about it.

I slid onto the stool next to Mac and asked Jim for a glass of their cheapest white, which is our joke about the very good quality wine they pour at special happy hour prices. I noticed that Sabrina was looking good, all in black, which is Tracy O'Grady's uniform for the bar staff. The wait-staff in the dining room wear white

shirts, which helps me sort out where I am after several stories and the accompanying beverages.

Peter patted me on the back as he went by to serve a table of smartly dressed professional women.

"So you never got around to telling me how you met President Eisenhower," I said, opening my notebook next to my right hand and producing a pen.

"Well," said Mac, "I met all the Five Star Generals or Flag Officers. Ike was special though. He *was* President, but not of the United States. He had been recalled from his position as the chief executive of Columbia University to become the first Supreme Commander of Allied Forces Europe, the new NATO command in Paris."

"There wasn't an issue with him being recalled from civilian life?" I asked, making a note.

"No. Five-star officers were never considered to have left

active duty. They retained military aides as long as they lived, as Admiral Nimitz did. Omar Bradley was the last of them alive. He died in 1981."

"Was there an issue with Eisenhower being on active duty when he was elected President?" I asked. It had never occurred to me that there was an issue in that regard.

"No, Ike resigned his commission to serve as President. JFK recalled him to "active" duty when he took office o replace him, with a date of rank restored to the day he resigned."

"Far out," I said. "So what was the reason for your visit? It was in Paris?"

Mac nodded. "Ike arrived in Paris on New Year's Day, 1951, and set up shop with his Planning Group at the Hotel Astoria. It was a nice location in the 8th Arrondissement, near the embassy and the Champs-Elysees and the Opera. Ike had a strange long office, with his desk in the middle. When we were shown in, he got up and walked toward us. He stuck out his hand and said "Eisenhower," which was a bit surreal."

"I imagine. He might have been the most famous American alive at that moment."

"Indeed he was, and that is why they wanted him to be the first NATO Military Commander. There were a lot of details to be worked out, but with the Berlin Airlift just concluded the year before, the new battle lines of the Cold War were in place."

"From Stettin in the Baltic to Trieste in the Adriatic an "iron curtain" has descended across the Continent," I said, quoting Churchill's words from the speech at Fulton, Missouri.

"And that is exactly why we were there to see him. We needed to get his permission to take a look behind the curtain. We could never do what we did without specific authorization."

"And what was that?" I asked, thinking about the kinds of special reconnaissance programs that existed in the world before whizzing satellites.

CINCNELM HQ, 7 NORTH AUDLEY STREET, LONDON, UK.

"Well, that is a story all in itself. I got to London in 1950, and picked up my car at Southampton and got Billie settled in a flat not far from 20 Grosvenor Place, where the Navy had their headquarters."

"I have been there a couple times," I said. "The Fleet Ocean Surveillance Facility Europe was there before the consolidation with the Royal Navy's facility at Northwood. We used the entrance on North Audley Street. You know the Navy moved out in 2009."

"Pity. We had the building for a dollar-a-year lease that dated from 1942 when Admiral Stark moved in to establish COMNAVEUR. Ike briefly had his headquarters there before D-Day."

IKE'S STATUE GAZING AT HIS FORMER HEADQUARTERS BUILDING.

"I have a picture of his statue in the square, placed so he is looking at the building."

"Better than looking at that ugly Embassy building. The Brits used to call it "Rooseveltplatz," because it was so ugly." Big Jim refreshed my wine, and Mac took a sip of ginger ale a little wistfully since the Quacks had restricted his alcohol consumption due to his medication. "Here was the situation. After the war, the primary

Navy mission in the occupied countries was to disarm the Germans and round up the war material.

There was a huge Scientific and Technical mission to assess the state of the art. There were some astonishing things in Hitler's bag of tricks. Hydrogen Peroxide-powered subs were just one of the breakthroughs that never made it into the war. Jet bombers. Ballistic missiles. All manner of things that we scooped up, along with the German engineers who had created them."

"Werner Von Braun and his team were collected under Operation Paperclip and relocated to Huntsville, Alabama," I said.

"Yes. And the Russians were vacuuming up Eastern Europe, too. All the harbors had been cleared of mines and debris and the geographic area of responsibility had been dramatically expanded. So, the command's name became was changed to Commander in Chief, U.S. Naval Forces, Eastern Atlantic and Mediterranean. That was the command I reported to. Admiral Robert Carney became CINCNELM in December 1950."

"So, what was it that Admiral Carney had to ask General Eisenhower's permission to do?" Mac smiled. "A thing called the Peacetime Aerial Reconnaissance Program."

"PARPRO," I said. "Flying around the periphery of the iron curtain to collect intelligence. That was around in my time, too."

"Well, sometimes around it," said Mac. "And sometimes right into it. But that is going to take another ginger ale. And what we told Ike helped mold his vision for what he was going to do as president."

"Which was?"

"Open skies," said Mac. "And in more ways than one."

"They tried to change the name of the program a few years ago. They wanted to call it 'PRCSO,' or Peacetime Reconnaissance and Sensitive Operations. It was an unfortunate acronym, since it came out "prick-so.""

Mac smiled. "They also called it "SENSINT," and you would be surprised at what went on."

I wrote it down in capital letters. "Tell me more, Sir." And so, he did.

NAVY PARPRO P2V2 MARINER OIL PAINTING.

"In those days, the military was still filled with combat veterans who understood how to successfully approach a target. We knew that the electromagnetic response to perceived threat could be mapped and plotted. Sometimes we would run attack profiles on Russian installations to see what would happen, breaking off before we broke the sovereign territory. Although it was peacetime, we felt we needed to be ready for war at any moment, and the more we knew about where the Soviet radars were located would enable successful strikes. If we needed to do it, that is." He looked across the bar to the well-stocked liquid reserve behind it.

"We flew specially configured aircraft along the Soviet littorals and those of their Warsaw Pact satellites to see what lit up and how their air defense command reacted. The dramatic developments around the successful penetration of the Manhattan Project made the interior of the Soviet Union a most important target, and there was talk of a growing "missile gap" in terms of capability between what would become the binary superpowers. And that led to one of Ike's biggest crises."

"You mean the U-2 shoot down in 1960?"

Mac nodded. "Not a good morning for anyone after that. I was back in the States by then, but the fact that the Soviets had developed a surface-to-air missile capable of reaching the U-2 at altitude changed everything."

"I read that Lee Harvey Oswald had been a radar operator assigned to Atsugi, Japan, where the missions originated. He would have been able to tell them just how high and fast the U-2s could go."

"That would be an interesting aspect of his defection to the Soviet Union in October of 1959. I have some thoughts about all that, and the Crown Jewels and LBJ. We can talk about that sometime."

I nodded, wondering if Mac knew something about one of the other great mysteries of the American Century. "Thankfully we had developed an on-orbit capability by then," Mac said. "And no one ever was able to bag one of the CIA OXCARTs. Those and the SR-71 Air Force sisters are the most amazing airplanes that ever flew."

Mac nodded in agreement and took a sip of ginger ale. "I am tired of the medication that makes me drink this stuff instead of beer," he said. "But like Ike, sometimes you just have to do what you have to do."

WITCH HUNTS

• • •

SENATOR JOE MCCARTHY IS ADVISED BY ATTORNEY
ROY COHN DURING SENATE HEARINGS.

There is a lot of talk about Russian interference in the U.S. election recently—the cyber hacks that CIA insists were a Russian job and the FBI which has a more cautious view on the ability to attribute the thefts to Moscow. It will be a political football for the incoming Administration (assuming it does) with the idea that Mr. Putin wanted Mr. Trump to win the election, and took action to muddy the electoral waters.

I don't know how that is going to turn out, but it is nothing new. The Soviets had been inside the United States government for decades, we still recall the witch-hunts that followed. Maybe we can have another one.

I remember talking with Mac at Willow sometime in March of 2012. There had been a flurry of news about an ancient Army counter-intelligence program called VENONA. We were trying to get him out when he felt up to it, and he is a rock star to the Regulars at the Amen Corner. I did not bother to take notes. My leg was bothering me still, and we talked about the future rather than the past, which included plans for the 70th anniversary celebration of the Battle of Midway, and whether or not we should wear tuxedos for the Ball (Mac said he wouldn't). He is at an age where he gets to establish the dress code, and we all supported him.

We had been talking about the great pivot in his career—the chance meeting with Admiral Forrest Sherman in the passageway at Main Navy which stood over on Constitution Avenue, now the site of the Vietnam Memorial.

Mac had come back to DC from Hawaii in 1946, and the transfer to the Restricted Line—Special Duty Intelligence occurred just as the Admiral had told him. He joked later that he never ran into the guys whose name was dropped from the list. He married Billie, and served for a year in the Pentagon and at Foggy Bottom as an editor on the Diplomatic Cable, a summary product of the War Department, Navy and State. That written product is the lineal antecedent of today's National Intelligence Daily, the famous 'NID.' I don't know if President Truman got a daily brief on its contents (the literature suggests the Arlington Hall VENONA decryptions were not passed to FDR while he was alive, nor to Truman, though one would think that at least the magnitude of the penetration of the Roosevelt Administration by Communists and Fellow Travelers would have contributed to Harry's muscular response to Uncle Joe Stalin in the great defense re-organization that culminated in the passage of the 1948 National Security Act.

Then there was the VENONA matter, which we touched on often on the way to understanding the immediate post-war challenges to the code-breakers in the Army and Navy, who would be brought together by President Truman in the Armed Forces Communications Agency, headquartered at Arlington Hall Station across the street from where I live. It was a spectacular breach of the Soviet clandestine communications security, brought about by the exigencies of the German advance on Moscow: the print plant that produced one-time-use keys had screwed up and some 35,000 duplicate key number pages. The pads were distributed to Soviet NKVD agents in North America, among others, and made their communications vulnerable to cryptanalysis.

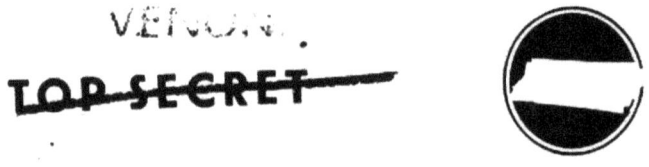

- 2 -   3/NBF/T1795 (Øf 12/7/66)

[i] KAL'MARO: Unidentified covername. This is presumably the same
                 word as French CALMAR, Spanish CALAMAR, Italian
                 CALAMARO, Russian KAL'MAR = SQUID. Possibly a
                 dialectal form. Also occurs as an addressee in
                 MOSCOW's Nos. 424 and 427 of 28th April 194Ø
                 (3/NBF/T1797 and 1798).

[ii] FINO: Unidentified. This could be
                 (a) a covername FINAUD = CRAFTY
                 (b) the common French surname FINOT.

1966 ANALYST NOTES ON A DECLASSIFIED VENONA INTERCEPT.

Nothing is easy, of course, and the body of the messages contained more code names for actual agents, but the stage was set for a serious controversy in the secret world. Army's Deputy Chief of Staff for Intelligence (G-2) Carter W. Clarke, a fierce anti-communist with no love for Uncle Joe and the sneaking suspicion that he might strike a separate peace with Hitler. Clarke initiated the program in 1943 to examine the cables sent in 1942 and the end of the war, when Human Intelligence about the compromise was received from an NKVD asset in the G-2 organization. The pads were recalled and replaced by the NKVD, but there was enough material for hundreds of analysts to work for four decades on the identities of the Russian Agents.

WILLIAM WEISBAND

Mac did not work on the project directly, but he had a knowledge of it from his time at Arlington Hall later, before moving the Navy element to Fort Meade. VENONA also set the stage and context for the Red Scare and the witch hunts to come, with a commie under every bed.

The slight problem was that the Soviets might not have been under every bed in Washington, but some of the ones they *did* hide under contained some very important people. Moscow had people in strategic places across the Agencies, and was tipped to the vulnerability fairly early on. The bulk of the successfully decrypted messages were harvested only up to 1945. An Army officer named Bill Weisband, a native Russian, was the mole who probably disclosed the success of the operation to his handlers in Moscow.

From 1941 to 1942, Weisband was an NKVD agent and handler of other moles in government and industry after service in North Africa and Italy. He was a gregarious fellow, and made fast friends in the community. Assigned to the Soviet Section at Arlington Hall in 1945, his fluent Russian made him valuable to both his ostensible and real employers. Weisband had access to all areas of Arlington Hall's Soviet work, including the Western atomic scientists who cooperated with the Soviets at least as early as 1944.

The Soviets apparently had monitored Arlington Hall's "Russian Section" since at least 1945. Weisband's earliest reports tipped off the Russians, and accordingly, Soviet authorities changed their diplomatic code and the VENONA Project decrypts dried up. He and Mac were posted to London at about the same time. Weisband never was accused of espionage, for fear that he would publicly disclose VENONA's existence.

The massive number of cables archived were still very useful and helped investigators to build evidence against the Rosenberg Atom Ring and British spies Klaus Fuchs, Donald McLean and Guy Burgess. The sensational trials that ensued were based on evidence and identity provided by the cryptologic source that had to be protected. The ambiguity of the code names in the VENONA documents made complete certainty on other suspected Soviet assets like Alger Hiss, organizer of the 1945 founding conference of the United Nations Conference in San Francisco, or Presidential Adviser Harry Hopkins problematic until the material was eventually released

The issue was an uncomfortable one for Progressive New Dealers who had supported our wartime Soviet ally and who had become embedded throughout the government during FDR's long tenure in office.

As professionals in the business, I do not have to delve too deeply into the Sources and Methods aspect of the story, and a lot of the bar-room conversation had a sort of cryptic wink-and-a-nod aspect to it. VENONA was a gift that kept on giving. Although only the years 1942-44 yielded much actionable data, the files were worked all the way up to 1980 before cryptologic resources were moved on to more pressing issues.

As professionals, we have all followed the remarkable opportunity for the analysts to cross-index the code-word agents with the Archives of the KGB in the early 1990s, smuggled out of Russia by former archivist Vasili Mitrokhin about his thirty years in the First Directorate Records Section.

That is the correlation between Alger Hiss and former Communist Whittaker Chambers in the Pumpkin papers and the Trial of Some Other Century. Enter Tail Gunner Joe, Dick Nixon and Ray Cohn the great witch-hunt, the real dimensions of which would have startled Old Joe, had he known the extent of all the networks. All the players in the witch hunts would take their turns on the stage as Mac was off having lunch with Tito (among others) during his assignment with CINCNELM in Naples and London.

ALGER HISS TESTIFIES IN THE PUMPKIN PAPERS TRIAL.

The ambiguity of guilt became a litmus test of political views across Our Fair City, since the very basis of the then-available evidence was likewise ambiguous. Senior army officers, in consultation with the FBI and CIA, made the decision to restrict knowledge of VENONA within the government. Army Chief of Staff Omar Bradley was concerned about the White House's history of leaking sensitive information, leading to the conclusion that senior aids to FDR were also Soviet sources.

The consensus today, after 40 years of analysis, is "yes." in the person of Harry Hopkins-and with Hiss at State and Harry Dexter White at Treasury, the Soviets had their bases thoroughly covered. Based on the news, I guess they still do.

Bradley decided to deny President Truman direct knowledge

of the project, and the "fact of" (absent the proof of the decrypts) in the CI and Intel summaries contributed to the . . . ambiguity. To some degree, Bradley's decision to keep the secret was counterproductive; Truman was distrustful enough of J. Edgar Hoover at FBI, and suspected the reports of Soviet penetration were exaggerated to obtain political leverage for the Bureau.

As we knew at the Willow Bar, Alger Hiss dined out on his persecution by McCarthy for decades, maintaining his innocence to the end. Only with the Moynihan Commission in the 1990s did the truth come out that Hiss was in fact identified as a Soviet agent through VENONA. In a crowning burst of rhetoric, hired legal gun Bill Kunstler made one of his last great quotes in 1995 (he died that year) when the Commission on Government Secrecy reported out. He claimed the Army had forged the whole VENONA project to discredit honest Americans.

I miss Senator Moynihan. He was a reliable Democrat, of course, but he was also a legislator who thought, and was not afraid of where the facts might lead his formidable intellect. We all have our limitations, though. Legislative insiders advised to try to do necessary business with the Senator before lunch on a working day.

I liked that version of Washington a lot better than today's edition, though it appears that some of the players are constant. At least it was a lot slower back then, and less daily evil could be accomplished.

MAC AND LIZ

• • •

Approaching from the west on Fairfax Drive, I saw Mac's champagne-colored Jag parked in the premier spot at the curb directly in front of Willow's patio. It had been raining all day, grim and gray and persistent, and the umbrellas were pulled down, somber, and water puddled the red paving bricks.

It was a good day to be indoors, and in a place that was nice and dry. I needed to get away from the roar of the industrial fan that had been blowing cold air to dry the seams in the parquet floors at

Big Pink since the flood. It was like living on the ship, working or sleeping near one of the great ventilators that forced air deep into the steel leviathan.

Unplugging the thing brought on a sudden silence that did not bring relief. Instead, it made me uneasy, just as it did on the ship when things went silent and that meant trouble.

There was no trouble at Willow, though. Old Jim was parked in his usual place at the corner of the bar, contemplating a cold long-neck Bud with earbuds from his iPod plugged into both ears.

I slapped him on the shoulder as I slid into the stool on the other side of the corner. "What did you do with Mac?" I asked. "His car is out front."

Jim scowled at me and unscrewed one of his earpieces. "What?" he said.

I repeated my question and he shrugged. "Don't know. I don't have the duty today."

"What are you listening to?" I asked. "You come to a bar to have some human interaction and then tune us out."

"I am listening to Joan Baez," he said with dignity. "And *you* just got here. Cool your jets." I looked up and saw the Admiral opening the outer door to the bar. He was wearing a bright red sweater and an amiable grin under his tan windbreaker.

"I wondered where you were," I asked. "I saw your car and thought you were here already."

"I was having a radio moment. Have you heard about the engine problems on the new mega-Airbus?"

"Between the cargo bombs from those jerks in Yemen and the Rolls-Royce engine problems, I am staying away from cargo jets and any airplane that has more than 400 seats."

"How goes the flood?" asked Jim, rolling up the cords to his earphones.

"The rug guys are coming tomorrow morning to pick up the waterlogged tribal," I said as I slid down a seat to let Mac sit between us. "Things may work out on that front. The plumber installed an aircraft-grade braided stainless feed to the ice maker to replace the plastic one that failed. "

Mac smiled. "The lesson, which we learn again and again, is to *never* go low-budget on things that handle water or electricity. The effects can be catastrophic."

I grimaced. "There are undoubtedly plastic connectors in the units above me. That's the hazard of an older building that has had significant and undisciplined modifications," I declared. "All of them at 'lowest cost, technically feasible,' if I can borrow from government language. And, of course, the phrase is nonsense."

Peter slid gracefully down the alley behind the bar with a sparkling tulip glass and a bottle of the Happy Hour loss-leader white. He knew what I would be having and did not have to ask. Mac leaned forward and said, "I will have a Bell's, Kalamazoo's finest."

I started to sing the lyrics to the old song as I saw foppish John-with-an-H enter the bar with a poker face. He stopped by Jim, handed something over, and then disappeared to his customary seat down the bar without a word.

"Did he pay up?" I asked. Jim smiled broadly. There had been a C-note on the outcome of the election in Nevada, and Jim was dead-on about Harry Reid, the fall of the House and the Democrat defense of the Senate.

"Damn straight," he said. "The man may be an idiot, but he is an honorable one. Unusual here in Washington."

I took a sip of white wine and felt my mood rising. That was accompanied by a glimpse of Elisabeth-with-an-S who was working the restaurant side of the bar this afternoon. She is-hate to say it-a willowy young woman with a graceful swan-like neck and auburn hair usually pulled back in a ponytail.

She is a graduate of one of the Case Western Reserve Law School, and she is bartending and working the tables just until she pays off the tuition bills, which she calculates will be by early 2032. "Hey, Elizabeth," I called out. "There is someone you need to meet."

I introduced her to Mac, who beamed with approval. He may be getting on in years, but he is still dapper and likes the ladies. He was proud of the new crimson sweater he purchased at Macy's that day to start the cool weather season. "Elisabeth was part of that crazy Halloween party. She wore a pink camouflage mini-dress with a matching fore- and-aft cap that was disturbing on several levels."

Mac smiled. "Pleased to meet you," he said, contemplating the image. "Likewise, I am sure," responded Elisabeth, sticking out her hand to take his.

"The Admiral is one of the last survivors of Fleet Admiral Nimitz's staff in World War two, and one of the architects of the victory at Midway."

"I was a code-breaker," said Mac. "They don't teach the history of Midway in the schools any more."

"I have heard about the battle," she said, smiling. I suspect she humors us just like Peter and Big Jim and Sabrina do. "I have to set up for the dinner service. I will be seeing you boys around the restaurant."

"The pleasure is ours," I said, suddenly remembering that there was a point to our meeting at Willow. I reached for my notebook and pull out my Pilot G-2 micro-fine gel pen. Serious business calls for serious tools.

"I wanted to talk to you about 1950, and why you were transferred to Naples and back to London, and the best job you ever had in the Navy."

Mac looked thoughtful. "That would be my time as a liaison officer at the British Admiralty with Nick. Nick Cheshire, that is. He was the greatest Russian naval analyst the Brits ever had. Spoke Russian, since his father had married a lovely Russian lady."

"How did you wind up at the Admiralty?" I asked, writing hastily.

Mac paused to let my pen keep up with his words. "Captain Ford was the N2. I was sent to relieve Ted Rifenburgh as the CINCNELM Current Intelligence Officer. Later in his career, Ted wound up commanding the Naval Investigative Service, but while I was en route to London with Billie, Ted managed to wrangle a six-month extension to line up for another set of orders."

"So there was no job for you when you arrived."

"Correct. Captain Ford decided he would rotate me through the elements in the Intelligence Division a couple weeks at a time.

Those were Current Intel with Ted, Technical Intelligence, Merchant Shipping, Political Intelligence and Admiralty Liaison."

"Which did you like best?" I asked.

COMMANDER LIONEL "BUSTER" CRABB IN GIBRALTAR.
PHOTO: IMPERIAL WAR MUSEUM

"Oh, Admiralty liaison beyond a doubt. That was one of the highlights of my career, working with Nick on the Russian Navy problem. Nick understood the Office of Naval Intelligence Y1 organization and our intelligence products. When I got to the Admiralty there were two other Americans embedded there doing merchant ship activities."

"That was a prototype for the modern mission, right? Like integrating Lloyds shipping data with operational reporting?"

"Close. Remember, we were working on five-by-eight index cards to keep our records. I got a chance to work direct with Nick on the Russian Current problem."

"That would have meant looking for intelligence on the new Russian Cruisers, right? Was Commander "Buster" Crabb murdered by the Russians while you were there? The Crabb Affair has never been solved for sure."

"No, that happened later, in 1956, when he tried the under-hull scuba swim when the *Sverdlov*-class cruiser *Ordzhonikidze* made a port call in Portsmith. But we certainly were hungry for any information we could get. The *Sverdlov*-class cruisers were the first post-war construction Soviet warships, and it was clear that Stalin was committed to supporting Admiral Gorshkov in building a world-class fleet."

SVERDLOV-CLASS CRUISER *SVERDLOV* UNDERWAY IN THE BLACK SEA. OFFICIAL PHOTO: U.S. NAVY

"Those must have been heady times," I said.

"Nick Cheshire wanted me to stay. But that was when Admiral Carney had to deal with the establishment of NATO, and the new structure of the Alliance in Europe. That is why we moved CINCNELM to Naples. I was picked to be Rudy Fabians's Deputy, and so off we went. Best tour in the Navy, though, working at the Admiralty."

"How did you get the family to Naples," I asked.

"We drove. But that is going to take another Bell's." The Admiral waved to Peter, and I took a pause to drink some of that marvelously crisp white wine.

REGIME CHANGE

• • •

Demonstration in Tehran, 1953, opposing Operation AJAX.
Photo: Public Radio International

"I told you the other day that I was sent to London in May of 1950 from DC." I nodded in agreement, and got out my pen and prepared to take notes. "I rented the flat, picked up our 1949 Mercury four-door sedan at Southhampton. We drove it to get around in London and then took it on the car ferry across the Channel to France and drove down to the Alps and the Italian frontier. Then, a year or so later, when we were in Naples and getting ready to go back to London, we had the start of the troubles with Iran that are still going on today."

"I was in the *Midway* battle group when they seized the U.S. Embassy in November of 1979, so I have been glaring at them in repressed anger my entire professional career," I said grimly. "Assholes."

"It goes back further than that, probably to about the time you were born," said Mac with a laugh. Big Jim deposited a ginger ale in front of him on the rich dark bar and a tulip glass half-filled with whatever Willow was pouring for happy hour white. It was always good, and I just had to specify the color.

IRANIAN PRIME MINISTER MOHAMMAD MOSSADEGH, 1951

"Here is how it started, as I recall. It caused quite a stir at the time. With the near-unanimous support of the *Majlis*, the Iranian parliament, Prime Minister Mohammad Mosaddegh nationalized the British-owned Anglo-Iranian Oil Company. We knew it as AIOC."

I took a deep swig of wine and said "They always seem to be taking things from somebody."

"Well, in fairness, the 1933 agreement under which AIOC was operating was widely regarded as exploitative and an infringement on Iran's sovereignty. That was in 1951."

"That was the British Empire, and it was the very year I was born," I said with wonder. "That is a long time to hold a grudge."

"Consider the grudge the Shias have for the Sunnis and vice versa for enduring hostility. With the British Services, we overthrew the Mosaddegh government and installed the Shah, who ruled for a quarter century."

"I recall that vividly, Sir. And the Iranians at Naval Air Station Pensacola who were being trained to fly the F-14 Tomcats and F-4 Phantoms we sold him."

"We were deployed to the Indian Ocean during all that," I said. "We all thought the Carter Administration demonstrated weakness in dealing with it. The overthrow of the Pahlavi dynasty in February 1979 made everyone in the region nervous. We wound up there twice in little more than a year. It was the start of the bad blood."

"Well, from the vantage point in Europe and Washington, the 1953 coup looked justified. But for many Iranians, the coup demonstrated duplicity by the United States. They called us hypocrites for presenting ourselves as defenders of liberty, but willing to use treachery to suit our own economic and strategic interests."

"If we don't who will?" I replied. "But is the Iranian reaction why Ike was soft on the Suez Crisis a few years later?"

"I don't know about that. For context, you have to remember that Iran's oil was the British government's single largest overseas investment. The Brits had played fast and loose with the terms of the concession that dated back to before the war. The Iranian workers who produced the crude were poorly paid and lived in slums. There was skullduggery, too. The Brits owned 51% of the company and bankrolled disruptive tribal elements and bribed officials to get what they wanted."

"The Iranians blamed Britain for most of its problems and public support for nationalization was passionate."

"Death to Britain," I said, raising a fist. Big Jim, the bartender, raised his right arm in solidarity.

"It was quite the affair. The Brits imposed a worldwide boycott of Iranian oil to pressure them economically, though the Attlee Government decided not to land troops to seize the refineries. With Churchill back at Number 10 Downing Street and Ike in the White House, opinion opposing a coup faded and they decided on what we call 'Regime Change' to oust Mossadegh."

SHAH OF SHAHS, REZA PAHLAVI

That must have been something to watch from London," I said, swirling the pale golden wine in my glass."

"Yes indeed. The coup was known as Operation AJAX, and required the Shah to dismiss Mossadegh from office. His family had to be bribed lavishly to get him to do it, but he came through in August of 1953 and got rid of him. The CIA's candidate, General Fazlollah Zahedi, was installed as Prime Minister "

"And they have hated us ever since," I said. "It is amazing it has been going on this long, and it looks like they will continue to export terror."

"I wouldn't be surprised in the slightest," said Mac, and the conversation meandered along through other times and places as it always did at the bar of the fabulous Willow restaurant.

"Under New Management"

I looked down at my notes spread across Willow's rich wood bar. The happy hour wine made me mellow, and Mac's stories were washing over me, not sepia-toned at all, but real, as if no time had passed at all since those people and those events occurred.

THE WHITE PALACE, BELGRADE, SERBIA. PHOTO: AP

"The Peacetime Aerial Reconnaissance Program—PARPRO— issue was part of what we did out of North Audley Street at CINCNELM in London," said Mac. "But it wasn't *all* of it by any means. There is all sorts of spooky stuff involved, and some activities that did not happen in the sky are still locked away in vaults."

"I know," I said. "I wonder of anyone still have those indoctrination sheets we signed, saying we would never reveal those programs under penalty of law?"

"If they still have any of mine, I would be surprised," he said, with a dismissive wave. "Paper disintegrates. Suffice it to say that Ike's feelings about the U2 program were shaped by his experience with the PARPRO missions in Europe, and the urgent necessity to assess what the Kremlin was up to with heavy bombers, rockets and atom weapons. As President, he supported all the technology that eventually made the U.S. over-flights of denied territory unnecessary."

"Too bad that Gary Francis Powers had to get bagged on the last U2 flight over the Soviet Union," I said.

ADMIRAL ROBERT BOSTWICK CARNEY, USN. OFFICIAL U.S. NAVY PHOTO.

Mac frowned. "Too true. But it was not supposed to be the last flight, though we were trying to stop. Tensions were high about the matter. And of course we were not in London for long. Admiral Carney became CINCNELM in December 1950, after I bounced around the staff for about six months. In June 1951, Carney assumed additional duty as Commander in Chief, Allied Forces, Southern Europe and CINCNELM Headquarters was moved from London to Naples to become CINCSOUTH. I had to find a new place for my family to live."

"Italy is funny," I laughed, thinking of the renowned hooker

Humpty Dumpty, who sat on the wall next to the road between the Naval Support Activity, Naples, and AFSOUTH HQ plying her trade with the Campfire Girls.

HUMPTY DUMPTY, THE NEAPOLITAN AMBASSADRESS OF GOODWILL

"Well, there is that. People said she and her comrades had been working that location since the Germans built the compound in WWII. But there were more strange customs. For example, when people moved out of rental quarters, they take the light fixtures out of the walls. I had to run around Naples to find replacements to mount so Billie and I could see. Did I tell you about having lunch with Tito?"

"You have mentioned it in passing, but how on earth did that happen? You were just a Lieutenant Commander, having lunch with the most powerful figure in the non-aligned world."

Mac smiled. "Special opportunity. It was December, 1951. Lieutenant Art Newel and I were sent from Naples to the northern Adriatic Sea. It was approaching Christmas, and we embarked in the Heavy Cruiser USS *Des Moines* (CA-134), Flagship of the

SIXTH Fleet. We were headed for a berth downtown in the splendid Adriatic port of Rijeka. We are riding a mountain of gray steel bristling with guns, and it is the first to visit Communist Yugoslavia since World War II."

USS *Des Moines* (CA-134). Official U.S. Navy photo.

"The people of the city are still getting used to the name Rijecka," I said. "Sort of like Ho Chi Minh City for the old Saigon. The Italians who had seized it at the Treaty of Rome in 1924—those that were still left—called it Fiume. When Tito's partisans arrived in 1945, 58,000 of the 66,000 Italian speakers fled the city, choosing exile to Communism."

"Summary executions of hundreds of alleged 'Fascists' followed the occupation, and there was a well of bitterness filled that only the strong man in Belgrade could keep from overflowing. The Croatians walked tall in the picturesque city, having thrown off decades of enforced Italianization."

"Rijeka then had the hallmarks of a great port, desired by all the powers. By turns, it had been Roman, Croatian, Hungarian, Yugoslav, Italian, German and then Yugoslav once more. It was

Tito's deepest incision into the European continent. An international force, including American doughboys, had even occupied the port briefly in 1919. Now we were back, eighteen hundred SIXTH Fleet sailors ready to go on liberty in a city that, until the day before, had been behind the Iron Curtain." Liz-with-an-S came by to top off my white wine and Mac smiled.

"There was plenty of potent slivovitz—plum brandy—waiting for our sailors ashore, and pretty girls and the other delights of the harbor that warm the hearts of all seafaring men. But of course, Art and I were *not* going on liberty. We waited impatiently for the brow to go across from the high gray hull down to the quay. We had an airplane to meet since the Major General who commanded Military Intelligence was coming by DC-3 to escort us to Belgrade."

VADM. MATTHIAS V. GARDNER ON BOARD SHIP AT NICE, FRANCE. CIRCA 1951-1952.

"See, the real meaning of this port call was a state luncheon with Marshall Josef Broz Tito, which would help broker a deal to try to help Tito balance the naval might of the SIXTH Fleet against the massive presence of the Red Army to the north and east."

"Vice Admiral Matthias Bennett Gardner, USN, was in command, and Art and I were hand-selected to provide intelligence support to the mission. Admiral Carney, CINCNELM in Naples, had authorized us to go with Gardner as his representatives, though he was confident that the SIXTH fleet commander could negotiate adroitly." I was scribbling furiously.

"Gardner was uniquely qualified in that regard. He was not only a naval officer but a naval aviator. The innate traits of each reinforced each other, and gave him the confidence to make big decisions without a lot of fuss."

"In 1945, while at a conference at a military conference at the Cairene Hotel in Egypt, he had selected the border between Russian and American-occupied Korea by gesturing at the 38th parallel. That matter was under armed discussion at the same time that *Des Moines* arrived in the harbor."

"Rijeka's airport is still awkward to get to even today, being located on an island adjacent to the city. There are distinct advantages to bringing your own boats to visit, and I highly recommend it if you have a ship large enough to carry one. It provides a lot of flexibility."

"I will remember that," I said taking a sip of wine. "But I think it is highly unlikely I will ever embark a gray hull again. Cruise ship, maybe, though I have never completely accepted the idea of going to sea for fun."

"The General's DC-3 swept down out of the gray skies and picked us up, quickly turning around for the flight to Belgrade."

"There were three days of talks in the capital, and we took up residence at the home of the American Legation United States Naval Attaché, or ALUSNA for short. He was a destroyerman by training. I will not mention his name, for reasons that will become plain enough, and he was an efficient and tightly-buttoned academy type. He was a prototypical Blackshoe, or ship-driver, just as Vice Admiral Gardener was an Airdale, or dauntless bird-man."

"Oil and water, or water and air, are those types. In those days, only two types of warriors earned special golden badges that proclaimed their specialties: submariners and aviators. The bubbleheads drove diesel subs and wore their golden dolphins with grim pride. They smelled bad when they got back—if they got back—from their dangerous undersea patrols. The Aviators wore the Wings of Gold, and smelled a lot better after an arrested landing on a pitching deck, provided they had not soiled themselves in fright." Mac smiled at the memory. "I started out as a Deck Officer, which is what they called Surface War as you know, and our blouses were unadorned with golden warfare devices. We were what the Navy did for a living, nothing particularly onerous, unless you consider dealing with high-pressure steam propulsion and high explosives an inconvenience."

"In Belgrade, once the sedans whisked us away from the airport, there was the official call on the U.S. Ambassador and the Chief of Mission, all of it leading up to the big lunch with the Marshal himself."

"The situation in Belgrade was tense, and the *Informbiro* crisis still reverberated as a threat to the regime. Tito was under intense pressure from Stalin to toe the Moscow line, and he was not going to do it. He was confronted with the threat of invasion, or assassination, and he desperately needed a card to play against the Kremlin. The U.S. Navy would provide him a jujitsu move, pitting the great continental land power against the undisputed ruler of the seas."

"Belgrade was a depressing place in winter, dark and chill, filled with an air of sorrow tinged with manic tendencies. Art and I did not even want to go out shopping, which is one of the great skills of sailors assigned to the Mediterranean Fleet."

"Like I said, we were billeted at the house of the ALUSNA. His long-suffering minion, the Assistant, was a Lieutenant named Mayo. He had to schlep the bags. Visiting delegations are the bane of overseas duty, and the more senior the members, the more stress. With VADM Gardener there was an emphasis on protocol, since he was the Fleet Commander, and Mayo's boss the ALUSNA were both thoroughly old school."

"The State Lunch was at the White Palace in the former royal compound in the exclusive Dedinje neighborhood. It was an imposing neo-Palladian pile, and is still so today if the pictures from the air campaign in the Balkans are to be believed. Famed architect Aleksandar Djordjevic designed it to the specification of the King and completed in 1936, though he never got to enjoy it. The King of the Croats, Serbs and Slovenes was murdered in 1934 while on a visit to Marseilles. The Queen and the children lived there until the war came, and her taste was reflected in the English Georgian and

19th-century Russian antiques she filled the place with. Tito took the place as his official residence when he took over, and he maintained the décor, which had been provided by the Jansen firm in Paris."

"It was good stuff. Jackie Kennedy used Jansen when she redecorated the White House in the Camelot days, if I recall properly."

"There was trouble in Camelot, from my experience, and there was trouble at the White Palace, too. As our sedans pulled up in the circular drive with the American diplomats and the naval officers in our Dress Blue uniforms, Tito's protocol officer nearly had a meltdown. VADM Gardner had a fat golden stripe and two smaller ones above it on his sleeves. The ALUSNA had four narrow gold stripes, indicating he was a Captain. I was wearing the two-and-a-half golden stripes of a Lieutenant Commander. There was a crisis of protocol."

"I can only imagine. I have seen those tempests in teapots before," I said underlining the name of the architect of the White Palace on the bar-napkin I was using to take notes. "What was this issue?"

"The burning question was this: was I too junior to dine at the same table with the Marshal of the Jugs?"

"The mission was in jeopardy, since the table had been set, and the slivovitz had been poured. The ALUSNA—the Naval attaché—was at his diplomatic best, though, and saved the day. "When I was a mere Lieutenant," he said primly, "I had dinner at the White House with *President* Roosevelt.""

"FDR had been elected President four times, and the protocol people knew that Tito had only been elected once, on a yes-or-no basis. The ALUSNA's declaration sealed the deal. I was permitted entrance to the vast oval table in the formal dining room. The delegation was carefully seated by seniority, alternating Jugs and Americans, and based on my junior status, I was astonished to find myself placed directly opposite the Marshal's empty chair, the best seat in the house."

"Precisely in keeping with protocol, Tito swept in, severe in his unadorned gray tunic, accompanied by his senior staff and translator. He was in his prime in 1951, handsome and chiseled, and still with a martial carriage that reeked of authority."

"You have to remember, this was the man who had faced down Uncle Joe Stalin, alive and in the flesh. It was pretty impressive. Introductions were made, and the toasting began. Slivovitz plum brandy to start, plum brandy with food and wine, and plum brandy toasts after lunch. Diplomacy is hard business."

MARSHALL JOSEF BROZ TITO, 1951

"I noticed that the Marshal seemed to speak perfectly good English, even if the formal conversation had to go through the herky-jerky of translation. The Marshal laughed at the punch lines to VADM Gardner's jokes before the translator could get to them."

"The Marshal had a key question, and all the ceremony on both sides was just the scaffolding to hold it up for consideration. "What can you do for me?" Tito asked, waiting for the words to bubble through the translator."

"VADM Gardner answered promptly, and with confidence. "We will send you an aircraft carrier, and put it Dubrovnik the second you need it. We will take you out on her, and show you flight operations. We will guarantee the security of the Adriatic.""

"Tito nodded. The matter was resolved, and the dining and jokes and toasting went on. Best Friends, forever. What do you call it on the internet? BFF?" I nodded and laughed at the notion, and Mac's knowledge of current social slang.

"Later in the Balkan afternoon, the gray sky was already darkening as the Marshal bade farewell to us and retired for a nap. The sedans pulled up and collected us for return to our billets at the ALUSNA's quarters on the hill."

"Chief Yeoman Quinley was the OPSCO, the operations coordinator, enjoying the shore duty, and with the luncheon with Tito a grand success, he ensured that a celebration was in order."

"The ALUSNA waved at the messman for whiskey, and our group settled down to discuss the future balance of power in the Eastern Mediterranean. We would be on an airplane back to Rejika in the morning, after all, and the heavy lifting was done. Only the reports remained to be written, and we juniors could handle that chore."

"The whiskey on top of all the slivovitz might not have been the best of ideas. Somewhere along the course of the strategic discussion the matter of special compensation came up. The ALUSNA announced that he thought the concept of "Flight Pay," a special bonus paid to aviators, was an affront to real Naval Officers, and should be immediately terminated."

"As a Naval Aviator, VADM Gardner, was naturally interested by the assertion. He grew more and more engaged as the ALUSNA warmed to his topic, eventually becoming quite fixated. As the level of the whiskey in his glass went down, the attaché's voice went up in volume and he went on ranting for an astonishing length of time. Eventually Admiral Gardner put his glass down and called for his car, saying he needed to call Washington."

"When he got to the Embassy, he actually made two calls. He told the Chief of Naval Operations that he had secured a deal with the Jugs that was going to poke Uncle Joe Stalin right in the eye. It was a triumph of naval diplomacy. Then he placed a call to BuPers, and told the Chief of Naval Personnel to get that son-of-a-bitch attaché the hell out of the country."

"When we arrived at the airport the next morning, we were a little under the weather, what with all the plum brandy and the whiskey on top. But not nearly as much as the ALUSNA, who had been directed by Washington to be on the plane that had already departed into the cloudy Balkan sky."

There were smiles all around, and we dug in our back pockets to get our wallets and get on with the rest of the evening.

When I got back from Willow I did some research about the

figures in Mac's tale, which is how these things worked. Mac was so kind with his time, and patient with the questions posed by his impertinent interlocutors that we could go over the stories as many times as necessary to get it straight.

Uncle Joe Stalin died on the fifth of March, 1953. He might have been poisoned, and he might not. His successor, Nikolai Khrushchev, once he was convinced that the monster was really and truly dead, denounced him. He reconciled with Tito in 1956, and the Marshal had a lively career as an independent and mostly benevolent despot thereafter.

He was a considered a Father figure by most Jugs, and they sung rousing songs about him, and every year on the Marshall's birthday, a child was selected to make a small speech, hand him flowers, or present the ceremonial *stafeta* at the end of a relay race."

"Of course there were problems, given the history of the region. But the Marshal maintained the semblance of unity by sending dissidents to work camps, or demoting them from positions of power.

With his death came the start of the horror of dissolution of the national agglomeration called Yugoslvia, created by the caprice of the Treaty of Versailles

Slobidon Milosovich lived in the White Palace for a while, but he is elsewhere now, and it has been given back to the Royal Family.

The ALUSNA in Belgrade was (eventually) rehabilitated, and continued a distinguished career. He served another attaché tour, this one in Moscow. He died in 2007, then the oldest living graduate of Annapolis. He outlasted VADM Gardner by many decades, but never had a kind word for aviators or their flight pay.

USS *Des Moines* was laid up long ago. The Navy considered her too expensive to operate. After years of disintegration in the yard at Philadelphia, she was considered as a candidate to be a memorial ship in Milwaukee, Wisconsin. The effort failed, since there was no public sentiment supporting the heavy cruiser's placement on the waterfront, and no apparent connection to the Badger State. She was cut up for scrap in Texas in 2008.

The U.S. SIXTH Fleet continued to call on the Jugs, mostly at Dubrovnik, on the Dalmatian Coast, for the next fifty years. My pal Chuck was there for several visits in the Nineties, before and after things fell apart. He saw the graceful medieval Mostar bridge that unified the Muslim and Orthodox Christian sides of town, when it was up, when it was down, and eventually when the graceful structure was reconstructed.

He says the real thing was a lot better, but then, you would expect that. He was a SIXTH Fleet sailor, and thus a most discerning tourist.

We were just happy that Mac was still going strong, stronger than I am. He was 91 that Fall, though only slowing down on the consumption of slivovitz, at least on weeknights.

THE HIGH LINE

• • •

USS *DES MOINES*, CA-134, IN A RISING GALE OIL PAINTING.

Peter is a real pro of a mixologist, and the heart of the spirit of Willow's bar. He snuck up in his unflappable way and filled my tulip glass to the precise level for easy listening to Mac's story of lunch with Marshall Josef Broz Tito, Emperor of the Balkans and the Dalmatian Coast.

"The weather was crappy," said Mac. "But the Jugs were determined to get Admiral Gardner and our party back from Belgrade to his flagship in order to get the heavy cruiser *Des Moines* underway from Rijeka on schedule."

"The port visit was scheduled for four days, and four days only. Tito's people were very strict about that. With the ALUSNA on his way back to Washington on the early flight to receive his dressing down, we boarded the plane and waited for a break in the clouds. The wind was freshening, and when we got airborne, it was a short but bumpy flight."

"If *Des Moines* left on the 23rd, she had plenty of time to get to her next appointed mission, which was a Christmas Day visit to Athens. The intent was to buck up the liberals, who were having their problems with the local Communists. 1952 looked to be an unsettled year in Greece, and the presence of the big warship was just the right statement about America's resolve."

Mac looked at his Virgin Mary a bit pensively and stirred the concoction with the celery stalk that rose from the red depths.

"Sedans were waiting at the airfield and whisked VADM Gardner and our party back to Des Moines with time to spare, and check the block for "Mission Complete." At least it was "MC" for him. Art Newel and I still had to get back to Naples in order to make the Holiday with our families." He took a sip of tomato juice. "Therein lies another tale.

"The notes of the meeting with Tito merited our expeditious handling in order to get them to Higher Headquarters. Admiral Gardner was once more insulated by the blue-tile linoleum of the Flag Spaces on the cruiser, and the unusual intimacy with us required by the mission was abruptly terminated once we crossed the Quarterdeck of he big ship. The icy remoteness of rank and command were once more imposed."

"I hear that," I said, taking a long swallow of pinot. "Traveling overseas with Congressmen, I actually came to believe they were real people. Unreasonable, but real."

Mac smiled. "Rank in the Navy, as you know, is a thing of wonder. We live in such forced proximity at sea that the barriers between us consist of vertical social stratification, from the mess decks north through Officers Country to the Flag spaces in the superstructure."

"*Des Moines* had her engineering plant on the line in preparation for getting underway, and we pulled out of the old Italian port that had become property of the Jugs, and proceeded into the teeth of a rising gale in somber colors: gray skies; gray ship, gray sky, gray water topped by white foam. Despite her bulk, the ship moved around briskly in response to the power of the storm."

"The warship was bound on important business for the United States; thus the need to get Art and me to the nearest friendly airfield at Trieste or further transportation to AFSOUTH Headquarters at Naples was only a tangential requirement that would be accommodated while underway to Athens."

"We never got to that port in our Med Cruise on *Forrestal*," I said. "I wish I could have gone to the Acropolis."

Mac nodded, having been just about everywhere. "So, once in international waters, the cruiser hoisted ball-diamond-ball on the mast to signal 'restricted maneuvering' and set a course as steady as she could in the heavy seas. A plucky destroyer came alongside in the swells, and the deck party prepared shot lines to go across and set up a Hi-line transfer."

"Oh man," I said, taking a sip of Happy Hour white. "Once, out of boredom and curiosity, I stationed myself in the background as the bridge team aligned my carrier *Midway* to come alongside a fast stores ship for underway replenishment. We were going to take on provisions, fuel and ammunition through VERTREP—vertical replenishment—by helicopter and by underway replenishment direct from the stores ship via lines stretched between the ships. It took miles to set up the position properly, and the consequences of failure were grave for the hapless young OOD."

A DESTROYER ALONGSIDE FOR REPLENISHMENT IN HEAVY SEAS.
OFFICIAL U.S. NAVY PHOTO

"That is one of the things we do well as a Navy, and the Chinese and Indians will have to get good at, if they are to be real Blue Water navies. Without replenishment, the great ship lose their military value swiftly. I will bet that your UNREP was in the gentle swells of the Pacific."

I nodded in agreement. "Smooth as silk and still hairy. I was standing behind one of the lines and a Chief yelled at me to get clear, since if the line parted, the bitter end would snap back and cut me in half."

"As we closed, and the destroyer they were going to hi-line us to fell into formation, I could see the front third of her full coming out of the water alongside. Art and I were directed to the weather deck. I could feel the crackling tension in the deck crew as they prepared the shot for the messenger line. Art and I were in our Service dress blues with the gold braid on our combination covers pulled down as chin-straps to keep them from blowing away in the gale. We had our little duffels clutched to our orange Kapok life vests."

"A rudder casualty or other navigational mischance would cause the ships to plow into one another, and there would be hell to pay, at least for the destroyer. Damage and lost careers at a minimum, death maybe, with me or Art in the middle."

"That day before Christmas Eve, only a hundred feet separated the two ships. The heavy cruiser at 17,000 tons displacement handled the rising seas well; the DD bobbed wildly, plunging in the waves. The process for us was the same as when the ships transferred inert mail and movies. First, the deck gang fired the guns with the messenger line attached.

"One or two tries might suffice to get a line across. Sometimes, more attempts were needed. Once across, and assuming no one on the receiving ship actually got hit, the light steel line was hauled in and secured, followed by another, heavier line and rigged with the sling in which we were supposed to ride. The changing tension on the wire caused it to rise and fall like a yo-yo, sometimes dipping into the waves far below."

"It was the only time I had to do it," he said. "Thank God. It was a terrifying and foam-flecked adventure. Done expeditiously," he said, "the evolution might take an hour in peacetime. There were those on the bridge of the cruiser who had done this in the war, when enemy aircraft might suddenly appear and they viewed this as an excellent opportunity for training the crew for emergency break-away. They hauled us with alacrity and a minimum of immersion, but it is not something you want to do while attired in Service Dress Blue."

"The breakaway, once the transfer was complete, was done with élan. Not to mention relief on the part of the bridge team that could then concentrate on navigation without the consideration of having *Des Moines* cut them in two. The skipper on the tin can then could move on to the next item on the Schedule of Events, which was a brief stop at Trieste for fuel and to disembark me and Art. All business, no liberty for the crew.

"We snagged a driver and a duty car at Fleet Landing, and made our way to the airport, where a Navy Air Transport Service C-47 *Dakota* was to pick us up. Despite the weather, NATS made gave it the old college try. We could hear the engines of the transport at the appointed minute of its arrival, though that was as close as it could get due to the thick clouds over the field. There was no instrument assistance at Trieste, and the plane circled above in the thin gray light. There was no hole to pick in the thick gray wool below the Dakota, and no way to land safely."

"The pilot was ordered to return to base in Napoli, sans PAX, and mission complete for him and his crew, if not for me and Art.

"For our part it was another ride in the duty car, this time to

306 ••• Vic Socotra

the train station in the old central city. The great waiting area was thronged with holiday travelers. At the ticket office we bargained in broken Italian and English for first class tickets to Roma, via Rapido, with onward transportation via the *Metropolitana* line to Naples. There was salt crusted on our uniform pants."

"Seated in first class, we realized we were exhausted after the adrenaline rush of the hi-line transfer and the confusion at the airport, dress blues wrinkled, the overnight journey did not seem so bad. The rains had been awful that December, and the North of Italy was flooded. Shortly out of the station, we had one or two snorts from the bottle of brandy Art had secured while I negotiated the tickets. Then the brakes failed on the only first-class car. The smell of burning asbestos pads filled the train, and with confusion and great show of energy, we eventually found ourselves in the only seats available, the hard wooden benches in the unheated Third-Class car."

"No amount of brandy could warm the seats, and the coffee we bought to doctor the liquor with was gone. Our blues, which had looked so trim and proud with Marshall Tito at the White Palace, were now wrinkled and our once-snowy white shirt collars were gray as the skies. We were looking pretty drab, only a day away from dining with the Ruler of all the Balkans."

"We shivered through the night, passing through Rapido, where a dining car was added to the train."

"An Italian businessman in a well-tailored car coat boarded the train there, headed for Roma, his mistress and his holiday, in that order. He looked us and took pity. "Ah," he said in gently rounded

English, "You are officers, and should not be in such conditions. Join me, and let us share this journey in a civilized manner! It is my gift to you!"

"In the dining car there was civilization aplenty: white tablecloths and heavy silver, gleaming china, steaming coffee and a breakfast that stretched elegantly into lunch, the waterlogged countryside in muted green rushing by, clickety-click."

"Eventually the ancient ruins of the massive aqueducts began to appear, marching toward the Imperial City, and soon enough we were on the platform, watching our elegant benefactor disappear into the holiday throng, his smartly tailored overcoat draped over his shoulders, Continental-style."

MT. VESUVIUS LOOMS ABOVE NAPLES.

"Another train, this one not so long, took us down to *bella Napoli*. The Bay was the color of gun-metal in the dying light, and Vesuvius loomed darkly under the low skies. When we pulled into the grand old pile of the train station, we climbed down off the train and walked out to the street, stiff, tired and relieved to be done with it."

"I told Art I would see him at the headquarters in a day or so, and secured a cab. I directed the driver to the apartment block where I had secured a genteel residence for my little family. It was a curious place up the stairs that had seen better days, but it was good enough for us."

"You said you had to purchase your own light and bathroom fixtures when you moved in?" I asked.

HUMPTY DUMPTY'S DAUGHTER, CIRCA 1986.
COPYRIGHT DAVID COLEMAN 2010.

"That was just part of the merry anarchy of Naples. Those were the days when the original Humpty Dumpty sat on the wall soliciting business outside the base at Bagnoli, on the way up the hill to AFSOUTH."

"She and her daughter were known to generations of sailors," I responded with a shudder. "I may not have seen Athens, but I did see her."

"Eventually the taxi wound its way through the traffic and pulled up in front of the apartment. I fished the last of my lira out of my wallet and paid off the cabbie. Then I trudged up the grand staircase to our place that looked down on the inner courtyard. I put my key in the lock. Inside, Billie had some holiday candles going and the place was warm. I sailed my combination cover across the room toward the chair near the coal fire. Sounds came from the kitchen that sounded like dinner, and a small voice could be heard yelling from the direction of the bedrooms.

"Billie," I said. "I'm home! Marshal Tito sends his regards!"

THE CLIP-ON BOW TIE

• • •

WILLOW BAR. MAC AND VIC NORMALLY SET UP CAMP AT THE TWO STOOLS
TO THE FRONT LEFT. PHOTO COURTESY TRACY O'GRADY.

M ac took what appeared to be a satisfying sip of his glass
of Bell's dark lager. "So, where was I? Billie and I went
to London in May of 1950. We rented a flat, picked up our 1949
Mercury four-door sedan at Southampton and had transportation,
so we could get around, even though the London public transportation was excellent."

"And instead of relieving Ted Rifenburgh as the Current Intel Officer on the CINCNELM Staff, you wound up with Nick Cheshire at the Admiralty," I said, glancing at my notes.

ANACOSTIA NAVAL STATION, CIRCA 1947. OFFICIAL U.S. NAVY PHOTO.

"Yep. Ted was in the first class at the Naval Intelligence School. He was an outstanding officer. Captain Eismarsh had founded the Naval Investigative Service—what you know as the NCIS today." Mac frowned. "I was offered orders there in 1948, but turned them down to work on the Cable in the Pentagon. Wyman Packard's book "A Century of Naval Intelligence" explains the politics of the establishment, which Admiral Ernie King supported to professionalize the business. Forrest Sherman did, too."

"Useful thing, professionalism in your Spook Corps," I said. "I heard Wyman got some stuff into that massive book that some people probably wanted to keep secret."

Mac nodded. "Wyman was a good man, and he wanted the whole story, as much as he could, to be documented. Eismarsh was a pompous guy, the senior academic in the Navy and he never let you forget it. He was a Harvard language professor ONI hired during the war, and after it was over Navy decided to establish and intelligence school after the war. The Naval Intelligence School was housed at Anacostia in a bunch of temporary World War Two buildings. Where the park is now."

"I think that is where my grandfather camped when he came here with the Bonus Army in 1932 and General MacArthur burned down his camp and ran the marchers out of town."

"I met the MacArthur once. He came to Guam for the morning meeting one time. He was another pompous guy, though Admiral Nimitz forbade us to ever say so."

"Yeah, you said you met all the five stars."

Willow's bar was filling up with regulars and strap hangers. Both the Mike's were there, Short and Long-haired versions propping up the Amen Corner end of the bar.

Peter and Jim were solicitous, filling up my tulip glass with a

crisp and impertinent white wine as I scrawled notes. Foppish John
had departed for dinner at the Lyon Hall, the new restaurant that
occupies the art modern façade where Dan Kane's trophy shop had
been located for years in Clarendon, across from the big Agency
facility where I used to work. He made a point of telling Peter he
would not be dining at Willow on the way out.

He used to work on the Hill, and never lets you forget it.

"They should have named the place "The Trophy Shop,"
said Old Jim.

"I don't think the feminists would appreciate that," said the
shorter Mike.

I looked up pensively. "So your two year tour in London was
actually a little over six months on the front end, ten months in Naples
with Admiral Carney to set up Allied Forces Southern Region as a
NATO command under General of the Army Ike Eisenhower, who
was in Paris as the Supreme Allied Commander." I had to raise my
voice to be heard over the din.

"Right," said Mac. "AFSOUTH. That was before the French
threw NATO out and the Headquarters moved to Brussels."

"Then you did the trip to visit Marshall Tito?" Mac nodded as
I ticked down the list in my notebook.

Dapper Jon-with-no-H arrived with his hand-tied bow tie and
blazer, looking like he had just come from a day at the races and
ordered a dirty martini.

"I love the way you tie that tie," I said to Jon. "I wear clip-ons.
It is a personal fashion statement. At my old company they had a
dress code that said you didn't have to wear a tie unless you were

with customers. When I joined my current outfit, they said we were professional, and wore a tie ever day. I said, "Fine."

"So what did you do?" asked Mac.

"I went on eBay and bought thirty vintage clip-on bow ties in very strange patterns. It is my way of sticking it to The Man."

"But aren't *you* The Man?" asked Jon, tugging on one side of his rep-striped tie.

AMBASSADOR GEORGE VENABLE ALLEN.
PHOTO: CORBIS IMAGES, BETTMAN ARCHIVE

Mac laughed, he being the one grown up in the group who had actually been The Man.

"That clip-on thing reminds me of the trip to see Marshall Tito. We wore mess-dress to the luncheon, the uniform with the little jackets and formal shirts, cufflinks and studs. Well, it turned out I had forgotten to pack the black tie. I was pretty frantic about trying to find one in Belgrade on short notice, but Art Newel suggested that I ask the Ambassador to borrow his. I was a little nervous approaching George V. Allen, but he was a North Carolina gentleman and very kind. We went up to his quarters and he rummaged around in a bureau and found one. He handed it to me, and I opened up the butterfly hinges to clip it on."

I looked at him expectantly, wondering what the point was. Mac smiled.

IRANIAN PRIME MINISTER MOHAMMAD MOSADDEGH. PHOTO: LIFE MAGAZINE

"The Ambassador's mouth fell open. He had owned the tie for years and had no idea that it folded open. He must have ruined dozens of shirts trying to get it on closed up."

"We have rocket scientists in the diplomatic corps," growled Old Jim.

I was starting to lose the thread of my narrative as Big Jim the bartender topped off my glass again. The notebook was getting a little blurry, but he had mentioned an epic trip across the Middle East while he was in Naples. "You mentioned the other day that there was that wild odyssey with Admiral Carney across his area of responsibility. You said you flew from

Naples through Turkey and Bahrain and Saudi Arabia and Egypt and Morocco."

Mac nodded again. "We were running the oil trap-lines where the Soviets were meddling. The Admiral wanted to get a sense of what was really happening on the ground over which he was supposed to be able to operate if the balloon went up." teetotalers who refused the whiskey after dining with the Saudis got sick. The Arabs never washed anything. That wasn't true in Iran, of course, but there was trouble brewing and we have not seen the end of that yet."

"He was overthrown by MI-6 and the CIA, right?"

"Well, Pahlavi had some help. But that happened in 1953, and I was back in Washington then, teaching at the Intelligence Schoolhouse on Naval Station Anacostia. But we could all see there was going to be trouble."

"End of the colonial age," I said, underlining the word "Shah of Iran."

"Not the way the Iranians look at it," said Mac. Then he finished his beer and gave us a thin smile. "Two is my limit, Boys, and I think I will be moving along. I can tell you more next time."

And he did. But in the meantime, the usual suspects at the Amen Corner closed ranks and saluted as Mac left the bar and walked slowly but resolutely toward his champagne-colored Jaguar.

THE USUAL SUSPECTS

• • •

FROM LEFT TO RIGHT: THE LOVELY BEA, OLD JIM,
JON-WITHOUT AND JOHN-WITH.

I was going to talk to you about a couple other things this morning, but then the news of Frank Sinatra's death spread, and I had to draft an obit about his life and times, and I put aside an analysis of the origins of the Occupy Wall Street movement.

It is all quite curious and worth some discussion about the usual suspects. There will be time for that—the coming cold weather will drive the kids off the pavement soon enough, and it may be that this is just a dry run for the big demonstrations that will happen in the run up to the election next year.

I have seen this movie before, as many of you who are a certain age, and remember it well, since most of us were participants in the last widespread street actions in the waning days of the Vietnam conflict.

Mostly for fun, which is why a lot of the kids are doing it now, I suspect. But there is time to get to that in the next thirteen months.

Thoughts about the American Family Party, a front group of ACORN and the SIEU, were driven from my mind as the other news spread. about the Iranian plot to assassinate the Saudi Ambassador here in DC. The plot involved bombing the Peking Gourmet Restaurant over on Route 7, which is where the family celebrated Mac's 92 birthday.

Peking Gourmet Restaurant—target of al Quds Force? The Bush Presidents dined, like Mac, behind the screen at the rear where the Saudi Ambassador would be seated. Photo PGR.

When the al Quds Force is planning on bombing restaurants that serve exquisite Peking Duck, this gets personal. That is, if this is not a "Wag the Dog" moment presaging lobbing some cruise missiles at Baby Food factories.

I don't know what to think at the moment, though I probably will once I am on the road to Michigan again to deal with the Raven Affair this Friday.

In the meantime, we wound up at Willow last night. That will come as no surprise, but we had a bunch of folks over at the office for a late meeting, which I avoided, and instead swung by The Madison to have my copy of the new book by Eliott Carlson signed by Admiral Mac.

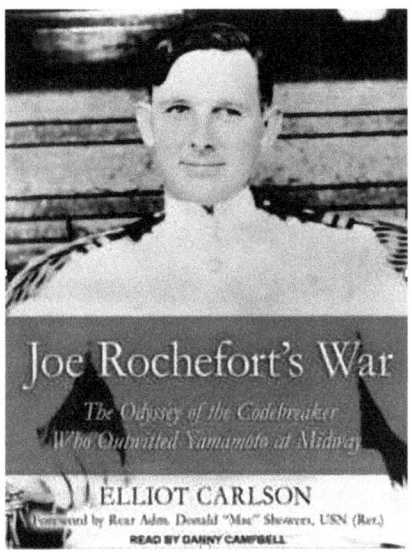

Joe Rochefort's War is one of those books that is absolutely definitive about a deep secret that proved to be the margin of victory in the titanic struggle of the Pacific War.

Mac wrote the introduction to the book, which I read in the advance copy in his apartment last week. Carlson had given him a couple copies in recognition of the more than forty interviews he contributed to the burgeoning narrative over the years.

"It is incredibly thorough," Mac said. "Carlson went everywhere to talk to the families of the men who have died, and to those few of us who are still around." I read him the words he had written as the forward to the book, and Mac smiled. "Those words are right from the heart," he said. "Now, you have to remember, this is embargoed until the 15th of this month. But take a copy with you to get started."

"I will replace it," I said, and slipped the thick volume into my briefcase.

I ordered the book from Amazon when I got home, figuring that it would probably show up a couple weeks after the embargo date.

I was amazed that Amazon paid no attention to the embargo date, and happily shipped it to me over the weekend. Accordingly, I decided to take his copy back and have him sign mine before heading over to Willow last night.

Mac is a little more frail than usual—it is either part of an inexorable process, or more likely, the insidious medication he has been prescribed for the slow cancer that all men will get, if we are lucky to live long enough. I read what he wrote on the flyleaf of the thick book, and thanked him for his friendship.

"I am sorry you can't make it to Willow tonight, Admiral."

"I know. Me too. If we can fix the medication maybe we will all be there again." "I sure hope so," I said. "I will stick my head in here before I leave for Michigan."

"Sounds good," he said, and picked up his magnifying glass to continue to check the footnotes. "And good luck with that. It is never easy."

I nodded, slipped the signed copy of the book in my briefcase and headed for the elevator, and the bright lights of Willow across the street.

The usual suspects were there, and more. I passed the book around, and Jim Champagne looked at the words that Mac had penned.

"He is an American Original," he said. "And not just one of the usual suspects." Then he ordered another Budweiser.

SOME OF THE OTHER UNUSUAL SUSPECTS, WITH WILLOW
FISH AND CHIPS. PHOTO: SOCOTRA

LET'S GET READY (FOR SCHOOL)

• • •

It was a Thursday afternoon at the Willow fine dining and drinking establishment on Fairfax Drive in the Ballston neighborhood of Arlington.

Big Jim and Liz-with-an-S and the irrepressible Jasper were attempting to manage the crowd at the Amen Corner at the street-end of the bar near the cocktail nook. Jon-without and John-with were enjoying drinks next to Old Jim, and Long Hair Mike was chatting up Jarhead Ray as they immersed themselves in the wonderful world of post-work at the Qualcomm building across the street.

I was meeting the Admiral for one of our weekly sessions, not that I wasn't there more than a few times a week. It was a cast of characters who had become, over the course of the decade, we occupied these stools, a family of sorts. We had got up to the point when Mac was recounting coming home with his family from Europe in 1953. The Korean conflict had raged while he was in Europe, and it was time for the military to adjust to the end of a war that had sucked five million Americans into the whirling vortex on the Peninsula.

I had two pens and remembered a notebook for a change. I got a crisp happy hour white from Liz-S, and opened the notebook to the last scrawled note halfway down the page.

"OK—back from Europe and back to the grind here in Washington." My pen hung above the notebook as I let some cool refreshing wine swirl on my palate from the tulip glass in the other hand.

Mac cleared his throat and prepared to resume the tale. "So, I had a total of about two and a half years with the Commander in Chief, Eastern Atlantic and Mediterranean (*CINCNELM*), a year of it in London and a year in Naples, then six months back in London before I returned to the U.S. in December or January of 1952-1953. I left in December and we arrived back here in January. When I left London to come back to Washington, I had orders to return to the Special Intelligence Section, Yl, and I was looking forward to it. I was obviously comfortable with operational intelligence (OPINTEL), and there were new threats to deal with as the Soviets became more truculent in their approach to diplomacy."

"Funny that we can look back on it with a certain fondness. But you missed the Korean War?"

"Yes, I did miss it. The war was over on the Peninsula in July of 1953, and my time overseas almost coincided with the entire duration of the conflict. So I never include it on my biography, even if the experience in the Police Action diverted attention from a growing Soviet threat, and framed many of the intelligence issues of the day. So, despite not having been there, Korea was a factor in everything: Rising Red China, Russian proxies and surrogates, all the things that were going to happen in the Cold War," said Mac.

"After the ship docked in New York, we drove down to Washington and I stopped to call the head of "Y" branch, who I think was then-Captain Bruce Weber and told him I was in the country and was going on leave, that I would be reporting in by such and such a date. He said, "I'm glad to hear from you; I'm glad to know all that, but your orders have been changed. You aren't going to be working for me; you're going to the Naval Intelligence School in Anacostia."

"My Grandfather knew the place in the 1930s," I said. "That is where he and his Bonus Army comrades camped when they were demanding Government benefits for their wartime service."

"Anacostia had not changed much. Still flat and green and prone to flooding. The school was located in one of the World War Two splinter-ville wooden buildings. I felt that the bottom had dropped out of my career. I had been rail-roaded out of the OPINTEL I loved so much. Instead, I was being sent to a back-water job that I had earlier turned down as being undesirable, and as far as I was concerned, was *still* undesirable. The only saving grace was that Captain Sam Frankel, a dear friend that I'd served under before at JICPOA in Hawaii. He was the Director, and I would have been happy to serve with him again in the right billet. I did call him up, and he was very gracious. He invited me and the family to stay at his home until we got settled, which wasn't necessary because we were going on leave to Billie's home down in southern Virginia. We got ourselves settled in, but I still was disappointed to not be going back into ONI and OPINTEL. Instead, I was going to be exiled to the wasteland of the basic school."

RADM SAMUEL FRANKEL. DURING HIS MILITARY CAREER, HE SERVED IN THE USS *TRENTON* (CL-11), *AUGUSTA* (CA-31), *CHESTER* (CA-27), AND *CHAUMONT* (AP-5). FOLLOWING THIS SEA DUTY, HE WAS A RUSSIAN LANGUAGE STUDENT IN RIGA, LATVIA, 1936-1938. HE THEN SERVED ON THE USS *ELLET* (DD-398) AS GUNNERY OFFICER AND XO. HE WAS ASSISTANT NAVAL ATTACHÉ, MOSCOW, AND ASSISTANT

NAVAL ATTACHÉ FOR AIR, MURMANSK-ARCHANGEL. IN THAT CONNECTION WITH LEND-LEASE SHIPMENTS TO THE SOVIET UNION, HE DIRECTED REPAIRS TO U.S. VESSELS, SALVAGING STRANDED AND ABANDONED SHIPS, AND SUPERVISED HOSPITALIZATION AND REPATRIATION OF SURVIVORS OF NAZI U-BOAT ATTACKS. LATER, HE ROSE TO BECOME DEPUTY DIRECTOR OF NAVAL INTELLIGENCE, AND HIS LAST JOB ON ACTIVE DUTY WAS AS CHIEF OF STAFF AT DIA, THE SAME JOB MAC HELD YEARS LATER.

"I can imagine the disappointment you felt after operating on major four and five-star staffs in real combat situations, or having lunch with Marshal Tito. The Schoolhouse would seem like a backwater."

"I made an appointment and told Captain Frankel of my misgivings, my concern and my general lack of interest in educational matters. He appreciated it and he was very accommodating. He said, "I'm going to give you six weeks to look at the curriculum and find out what we're doing and then I'm going to let you pick the part that you'd like to be responsible for. So I did that. I sat in the lectures and looked at the curriculum and it became evident to me, and fairly soon, that despite the changes made by Captain Hindmarsh and Captain Layton, and the good work that Captain Frankel was doing, the program was still weak on OPINTEL, or at least the way I understood it from wartime operations. There were limitations on what we could teach, since we could not talk about Communications Intelligence, which was still heavily classified.

"COMINT," I said. "But the Russians knew all about it. They had intelligence collectors everywhere in the government, and penetrated the Manhattan Project and just about everything else we were doing."

He nodded, still passionate about the times and his tour. "The Russians were superb in their Human Intelligence capabilities. Still are, for that matter. Back then, our students were not cleared. I don't believe there were any clearances provided for the school staff, and certainly not among the students, most of whom were brand new to the Navy. So you couldn't talk about COMINT as being a major source of what we called OPINTEL in the Navy. Instead, we had to talk about the other sources, so it made it a very artificial to try to emphasize the importance and significance, and the role of OPINTEL in supporting fleet activities and fleet operations in command decisions when you couldn't talk about the main ingredient of intelligence."

"By the time I went through the basic course, they had decided to send us all through the Armed Forces Air Intelligence Training Center at Lowry Air Force Base in Denver. It was a weird relic of Vietnam and the Cold War. We had to be in class at 0600 in the morning and they let us go at lunchtime, just as if there was another class coming in behind us in the Afternoon, which there hadn't been since Saigon fell. Was there ever any thought to establishing a sub-curriculum at the school for people who were heading for OPINTEL billets?"

"Not at that time. Of course, many years later all these things developed more realistically, particularly when they merged the All-source and Air Intelligence communities together in 1968. I think, now, students in many of our schools are cleared for COMINT and other compartmented subjects. At that time, in 1953, there were no clearances for instructors or students."

"At Denver, they worked our tickets in parallel with the course, and it was a big deal when we were getting to the Strategic block that

they gave us our Top Secret Special Compartmented Intelligence
Clearances. It was pretty cool. We thought we were real hot-shots.
I am sure you could have done a lot more if the students had been
cleared. But the school was really concerned with giving everybody
a common baseline of capabilities, right?"

"That's right, but it turns out that what was needed was even
more simple than that. What I was trying to do was identify the
areas where I felt that the curriculum was weak, and to contribute
to the development of Operational Intelligence. I found two main
areas of concern. One was naval communications, and the other was
the ability to write succinct messages and to handle them properly,
Naval Communications traffic was not a subject that was taught
at the school at all. This was important because, at that time, the
Naval Intelligence School was running two six-month classes per
year, and eighty percent of our input were from Officer Candidate
School and were mostly Ensigns."

"Just like you were when you reported to Station HYPO," I said,
laughing. "What was your throughput? "How big were the classes?
What was the through-put?" I asked, thinking of my little class of
twenty or so students at the Armed Forces Air Intelligence Training
Center at Lowry AFB in Denver.

"Our classes averaged 80 or so officers per class. In my two
years, I was privileged to work with and instruct perhaps 300 or
more officers."

I whistled at the number.

"Probably half or two thirds of those officers were Ensigns who
had been commissioned in the Officer Candidate School at Newport

and were now coming through Intelligence School. The balance of the class was composed of unrestricted line officers who were taking on intelligence as a subspecialty, and they ranged in rank from Lieutenant to Commander. Each faculty member was assigned a proportional number of students for whom he was class advisor.' So, in addition to working with the class as a whole, each instructor worked intimately with perhaps fifteen or so individual students for whom he was faculty advisor."

NAVAL INTELLIGENCE SCHOOL AT ANACOSTIA, CIRCA 1960.

"How skilled were hthey?" I asked. Mac had the grace to humor me. "Our students had just come into the Navy; they'd been through OCS; and now they were coming to learn about intelligence, and among other things they needed to know about communications, and the very basic ingredient of OPINTEL."

"They didn't know anything about naval communications either. Not how it worked, or how you wrote messages to effectively much less explain complex issues in a briefing format. The other thing I inserted into the course was what I called "observation and reporting.' "

"Useful topics for people who are supposed to be collectors of information," I remarked. "They used to work us over on sample briefings on Soviet military capabilities to the class at Denver. But the good news was that any classified research had to be done at the library, which was cleared for Secret. We couldn't take anything home for study and after a while, some of the people in my class would just go to the Officer's Club at lunch and stay until Happy Hour."

"I see some things haven't changed. We couldn't do that due to the clearance problem. What we came up with was a work-around that was useful to show the Ensigns how to do basic unclassified research. But we liked a hands-on approach. We used to take the school on field tri ps. One destination was the base at Quantico where the Marines put on some magnificent demonstrations of amphibious warfare capabilities. We also took them to Patuxent River to let them see the Naval Air Test Center, and to the David Taylor Model Basin where they test hull designs. A lot of the installations around here that show the exotic function of naval warfare, which we felt was a necessary part of the basic education in intelligence. We also encouraged them to look at the field trips the way a Soviet agent would. We wanted the students to observe and report, and they had to actually write on some aspect of something they observed in the course of these trips."

"The Navy had figured that out by my time to be educated in the mysteries. They sent all of us new guys from the schoolhouse to aviation squadrons to cut our teeth on how the strike and pow-er-project mission was really accomplished. We learned more about our Navy than the Soviets in that tour."

"Not much different that reporting to HYPO from Public Affairs school in Seattle," laughed Mac. "We'd get these intelligence reports in from the field trips, and I would have to review and grade them.

That led to the realization that most of the Ensigns of the 1950s who were coming out of college and into the Navy couldn't spell, and they didn't know English grammar.

"Still a problem today," I muttered. "Only it might be worse."

"You better believe it. When we went to Patuxent, you should have seen the variety of ways the word "helicopter" was spelled. It was just absolutely unbelievable. It was so bad that it led me to start a course in military terminology, spelling, and grammar. It actually was a course in collecting and writing intelligence reports. The librarian at the Intelligence School told me that, under the terms of reference to the school, I could not write a manual on basic English grammar and spelling. What we did had to directly be related to our profession."

You would think that effective written and verbal communicating would the main skills required in the intelligence business."

"I could not agree more. The Main Battery of intelligence, in fact. So, I wrote a little manual that we reproduced in pamphlet form called 'Writing for Intelligence Officers,' or something like that. But it was purely an effort to try to teach these college graduates—even the ones from the Ivy League schools—the English language. Those were the kinds of things Captain Frankel let me do. He was a good guy. He made flag, by the way, so his work was rewarded."

"Making Flag was never anything I had to worry about," I said. "Too much baggage. But I would be interested in finding a copy of the pamphlet, since the problem has not changed one iota since 1953." We both shook our heads at the sad state of public education. I looked up the Willow Bar, and things were just transitioning from the happy hour rush to the crowd that was going to be there for the long haul of evening. I sighed. That was not going to be me tonight, and we settled up our tabs with Big Jim, and I walked the Admiral out to his champagne-gold Jaguar.

"Next time, we are going to finish off the 1950s," I said hopefully.

"You might be careful with that. There was a lot of stuff going on then. It may take you a few more glasses of happy hour white than you anticipate." Then he climbed into his Jag and motored off sedately down Fairfax Drive.

TRADECRAFT

• • •

1966-VINTAGE SHRAPNEL FROM A ROCKET ATTACK IN SAIGON.

Willow was quiet, but it was a wary quiet, awaiting the arrival of people who would drink away the stress of the day. Old Jim was listening to music on his MP3 Player, Mac was dressed casually in an aloha shirt and slacks. Jasper was holding down the bar, waiting on Big Jim to arrive to handle the Happy Hour crew. Heather 2 was on the wait staff, though they had been auditioning her to work the bar as a fill-in, due to the loss of Briana-saurus Rex, who had left Willow for what we understood were "creative differences" with management about the cut of her tight Lycra top stretched over her imposing and legendary *décolletage*.

There is always a moment of silence when one of the bartenders departs the family, and Briana had been warmly welcomed, at least at our end of the Willow bar.

RADM DONALD "MAC" SHOWERS IN HIS CUSTOMARY ALOHA SHIRT.

I was in my grown-up work clothes with a jaunty bow tie. The Admiral fished in the front pocket of his slacks and brought out two objects. One was one I recognized immediately—it was a battered Zippo lighter with the squadron logo of Fleet Aerial Reconnaissance Squadron ONE, the West Coast intelligence collection aviation unit. The other was something I did not recognize—a nasty looking twisted piece of metal that despite its tarnished exterior radiated a passive hostility.

"OK, I'll bite. I recognize the lighter and have dozens of them. I used to collect them, in fact. But what it the other thing?"

"Well, you can add the Zippo to your collection, and we are going to talk about VQ-1 one of these afternoons. The other thing

I thought you might find interesting. I found it the other days going through an old box. It is a piece of shrapnel from a VC rocket that was used to attack the Meyercord Bachelor Officer's Quarters one night in Saigon while I was there. I walked through the wreckage the next morning and picked it up as a souvenir."

I picked it up from the bar. It was as ugly a thing as I have seen. The edges were still razor sharp where the high explosives that the steel shell casing had enclosed had turned it white hot, twisting and ripping as the missile tore into the cinder-block façade of the BOQ.

"That would get your attention right quick," I sad, putting it down before I nicked my finger on the sharp edges.

"You bet. Particularly traveling at more than a thousand feet a second. No one got killed in that attack. Not any of our guys, anyway. But that was the Saigon experience."

"I only visited that town once, Admiral. Much later, of course. But it was an amazing trip."

"Oh," he said, brightening a bit and leaving Saigon behind. Or ahead, if you were trying to follow the wandering Willow narrative. "After our talk last week, I was thinking about my tour at the Schoolhouse. I look back at it—the tour I didn't want to do, and upon reflection, it was one of the best and most satisfying of my Navy career. After talking about it with you last week, I thought of some other memories of that time. I was almost discouraged about having had my orders changed and having been sent to an environment with which I was totally unfamiliar. I hadn't asked for it, hadn't thought about it, and didn't really care much about it."

"What was the course work like? What did you try to teach

the new Ensigns if you couldn't use classified information to teach them?" I decided I really liked this version of the Happy Hour white. Curiously refreshing, and I almost lost track of educational challenges in the Intelligence Community. Mac was able to drag me back.

"Well, that was one of the challenges. "In the course of the two years that I was there, I instructed four classes, four complete classes, start to finish."

"I remember the courses at Denver started out with ship recognition and some basic photo interpretation. It was fairly basic stuff until our clearances came through and the instructors could talk more freely about what we were really going to be doing."

Mac nodded. "In my time the basic course started out with Professor Francis 'Frank' Decelles. He was, incidentally with the Naval Postgraduate School system, which had jurisdiction over us, He was the senior professor in the Naval Postgraduate School system. He was one of our links to Monterey. During our six-month course, would hold forth for the first six weeks full-time. All lectures. Essentially, he really lectured for about four hours in the morning. In the afternoon, then, we would have discussions, seminars, and various other activities to give some variety. But the first six weeks were devoted to Professor Decelles lecturing on world politics and the political scene, history, geography, whatever."

"It is a wide world out there and a lot of folks don't pay much attention until they actually have to visit it."

"I never did take the Decelles course, so I can't tell you exactly what he covered. I used to sit in on some of his lectures. He was a good lecturer; a flamboyant man who just *loved* to talk. He did this over and over, course after course. When the curriculum was expanded from six months to twelve, Frank's courses simply expanded to eight weeks instead of six, so he could talk more."

"I have shared the misery of having to listen to military people who thought they had a lot to say, and plenty of PowerPoint slides to back it up," I said, wincing at the memory.

"It can be painful. Then, after Decelles' course was finished, we started our instruction in various aspects of naval intelligence techniques and tradecraft and so forth. Field training was essential. As I mentioned last time, we'd go to Quantico, or Patuxent River, or David Taylor Model Basin, and we'd put them in the role of an attaché—a foreign attaché—observing a U.S. base or activity as an exercise in tradecraft. They would even take pictures, like the Moscow attaches would on Schmidt's Embankment in Leningrad, the only place we could get a look inside the doors of the covered building halls where the new generations of Soviet submarines were being constructed. Satellites could not see through the roof, after all. Sometimes, low-tech is the only way to collect effectively. After we returned from the trip, the students had to write a report. They would try certain little techniques of discreet activities not to get "caught" collecting. Some worked well, and some didn't. I think that's something that an individual designs himself and picks up rather than gets taught."

"I completely agree. When was your tour up at the school?" I asked, underlining the word "tradecraft" on the napkin in front of me.

"January, 1957, two years almost on the nose. We had just gradu-ated a six-month course before Christmas. I am glad we talked about it. In thinking back, and I still look back on it to this day, it was one of the most rewarding tours that I ever had in Naval Intelligence. I think it was mainly because I was working with a large number of people, and getting acquainted with them. Wherever I went and whatever I did for years afterward, I would encounter these people in various pursuits, assignments, and walks of life. They didn't all stay in the Navy, obviously. They would remind me that I was one of their instructors in Naval Intelligence School, and frequently they would remember, and I wouldn't—mainly because there were so many students and I didn't get that close to all of them. "

"I had the same thing happen after I was the junior Assignments Officer. I always wondered if one of my former clients find me in an alley or a dark parking lot and try to beat the crap out of me."

"Apparently I did a better job at the Schoolhouse," he said with that twinkle in his eye. "It happened to me all over the world for many years, and it still is rewarding that we were able to make an impact on so many lives. An example of that is Senator Dick Lugar.

He went to OCS and came to us as an Ensign; he did his time in the Navy and then went back to Indiana politics, became mayor of Indianapolis, and whatever, House of Representatives, and then a United States Senator. Many years later when I was with the Director of Central Intelligence, I was testifying before the Senate Intelligence Committee, and Senator Lugar came up to me and reminded me I had been one of his instructors at the Naval Intelligence School. He was not one of the students for whom I was faculty advisor, and frankly I didn't remember him being there. But that's one of the gratifying things that grew out of those orders to Anacostia. And we might have taught those officers how to read and write, which certainly worked out for Dick."

A YOUNG SENATOR RICHARD LUGAR, R-IN.

"As a general proposition, I believe in literacy for our public servants, Admiral." I took a deep pull off my Happy Hour White, put it back down on the bar and tried to remember what I had thought we were going to talk about. "But weren't we going to talk about your time at FIRST Fleet out in San Diego? That was my favorite place to live in all the United States."

"In those days I would have tapped a cigarette out of the pack and flipped open my Zippo to light it."

"I remember those days," I said, having had a Marlboro on the walk over from the office. "I guess the world has moved on a bit, wouldn't you say? Now, about FIRST Fleet . . .."

THE FIRST FLEET

• • •

COMNAVAIRPAC HEADQUARTERS, NAVAL AIR STATION,
NORTH ISLAND, CORONADO, CA.

B oomer was behind the Willow bar, and she was pouring with abandon. I had not had a good day at the office, and it was a relief to her about what her kids were up to, how her Dad was doing, and the rest of the maelstrom that is her life as a single Mom and breadwinner for her family. And living down by the Occoquan River, and don't get me started on *that*.

Mac Showers was dressed in a jacket and bright shirt and looked comfortable despite the sweltering heat outside. I was completely unprepared, which made me feel vaguely unprofessional, a sensation that I hate.

"So you picked up your family and moved to the West Coast?"

"We did, indeed. We went to San Diego and settled in Coronado because the headquarters then was in a building on North Island Naval Air Station. I relieved Captain Wyman Packard, the first of two or three times when I relieved him.

"It was a small Naval Intelligence community then, Sir."

CAPT WYMAN PACKARD, LATER THE HISTORIAN OF
NAVAL INTELLIGENCE, AS A MIDSHIPMAN.

"It certainly was. The most interesting aspect of going into the tour was that at the time I got my orders, Wyman and some of the staff, including the admiral, were deployed. That was the first, and I believe the only time, that Commander, FIRST Fleet ever was deployed in an operational role. They were gone for several months. I think they deployed in mid-1956 and did not get back until the end of the year. They were deployed to Southeast Asia because of the French problems in Vietnam."

"No kidding. You could have been part of the first American

intervention in Vietnam!" Mac smiled and nodded. "I suppose that is true," he said. "I always like to be ahead of the game when I can. That was at the time of Dien Bien Phu, after which the French pulled out from Indo-China. But there was the concern that the French, because of their problems, might request U.S. assistance, so the U.S. Navy was prepared to set up   a Southeast Asia Command that would have been a joint staff with Army and Air Force officers assigned. And to be ready for that contingency, FIRST Fleet went forward with part of his staff."

"But what forces did he have to influence anything?"

A BAD DAY FOR THE FRENCH AT DIEN BIEN PHU.

"I don't know the details since I was a non-participant, and this is just what I got from Wyman when we turned over. I understood that the Admiral borrowed some ships from SEVENTH Fleet. They had a couple carriers and supporting ships that operated in the South China Sea and were prepared to assist the French if they asked for help."

"There has been some controversy over that. The accounts I read suggested that the weather was too crappy to provide close air support."

"Probably true, but in the end, the French decided to abandon their empire and went home. They had problems much closer to home in Algeria, which was considered a part of metropolitan France at the time. That was far more emotional than far-off Indo-China."

"Funny you say that, Admiral. I was in Tokyo and drinking at the greatest bar in town at the Old Sanno Hotel, the place the military appropriated during the Occupation. You could see just about any old Asia Hand in there at one time or another. Anyway, I was sitting next to a guy a little older than myself and we got to talking. He had been Special Forces, and was interested in the military history of the French and American wars. He told me that in 1969 he had arranged for an insertion mission at Dien Bien Phu so he could see it."

"I guess that is what you call a combat tour!" laughed Mac. "I suppose I am not surprised, since all the regular Vietnamese forces were pulled south to replace all the VC who were killed in the Tet Offensive the year before."

"It certainly blew my mind, but another pal who was in the Special Operations Group said they went north of the DMZ frequently and it was safer up there if you minded your own business."

"Well, that opens up a couple cans of worms, but that takes us forward to Vietnam and I thought you wanted to talk about FIRST Fleet."

"Well, my interest stems partly from the fact that I was on the THIRD Fleet Staff twice, with radically different missions. The first time, in Hawaii, we were ashore at Ford Island and supposed to be the Theater Anti-submarine Warfare Commander. Our main mission was targeting Soviet boomers. The second time we were in San Diego like you were, but although we had a flagship, we mostly did training and certification for deploying Strike Groups."

"I will beg ignorance of what the THIRD Fleet people are doing now, but we were primarily responsible for the same thing: training and readiness among type-commanders who were then resident on the West Coast."

"Same deal for us. AIRPAC and SURFPAC are still there. They used to each have an intelligence Captain billet, and some of our crusty old guys would trade jobs every three years because they liked living in San Diego and wanted to stay there. We called it the Senior Circuit."

"I was a Commander then, but we had a lot more opportunity for officers who wanted to stay away from Washington. In addition to the Airdales and the Black Shoe Type Commanders, we had the Cruiser-Destroyer people, the Amphibs, and even SUBPAC and Service Forces had reps there from Hawaii where they were headquartered."

"Wow, I had no idea they had actually streamlined things. It was a bewildering mass of competing equities even in my time."

"Under guidance from PACFLT HQ, we were responsible for herding the cats through unit, type and inter-type training of the non-deployed Fleet forces, preparing them for contingencies, emergencies and the deployment rotation to the SEVENTH Fleet AOR."

"There are some constants in the Navy," I said, taking a sip of vodka tonic. Mac looked at me with a questioning glance. "I know. I normally don't drink hard liquor when I am out for Happy Hour at Willow, but it was a bad day at the office. Not as bad as a Comprehensive Training Exercise off Camp Pendleton, but you know what I mean."

VADM Sol Phillips in Command.

"FIRST Fleet was a great tour, great quality of life, but as you probably know better than most, there isn't a whole heck of a lot of intelligence going on. Our division was a very small office, The N-2 had one assistant and a yeoman, and that was it. The Assistant N-2 was normally an aviator and a sub-specialist. As one of my collateral duties, I was the Special Security Officer—the assistant normally was not cleared for COMINT. Those were the days when COMINT clearances were tightly controlled and extended only to a few people. I had a large double-door safe in my office that I inherited from Wyman. I kept all the COMINT materials in there, the studies, messages and background papers. I was the sole custodian, and I had to open up in the morning and close it at night, and remember to lock it up if I was going to lunch or make a head call. The only ones on the staff who were cleared were me, the communications officer, the Admiral and the Chief of Staff. Ops, Deputy Ops and the Plans guy."

"So, it sounds like a great tour to be with the family in San Diego. It sure was for us."

"Oh my, yes. It was marvelous. Coronado Island is one of the great places on earth. But I should tell you about our Admiral—"Sol" Phillips. He was one of the great characters I knew in the Service."

It looked like Mac was ready to tell the story, but I was confronting the bottom of my glass and decided that the three of them had been plenty since Boomer does a great pour and I had to drive. "I will catch that one next time, Sir. I am just amazed that our experiences were so similar in times so far apart."

"I don't think the missions were any different for any modern Navy," said Mac. "You fight like you train, and that is what sets apart a first-class Navy from all the rest."

"Amen, Sir."

OLD NAVY

• • •

Mac did not have a good weekend, nothing particularly bad, but the Docs had taken beer away from him again. Boomer had the shift off to do family business, and Big Jim was covering for her along with Liz-with-an-S. There was energy in town, the elections and the partisans on both sides had become irritating. It looked like a time to move forward, perhaps to Mac Shower's time as the Pacific Fleet Intelligence Officer, the job once held by his old Boss, Eddie Layton. That was not going to happen. Perhaps the Virgin Mary that Big Jim assembled for him was not exactly what he wanted, or something else had driven a mildly pensive mood to the afternoon. I was fairly upbeat, since the day had gone as well as could be expected at the office, and we had actually won a task order from the government on the big open-ended contract I allegedly manage for the company.

I was having the Happy Hour white, and I expected to have a few of them while we talked. "So, on to the Pacific and Vietnam," I said, lifting my pen.

"Not so fast. You got me thinking about the last genuine character in the U.S. Navy: Vice Admiral William Kearney Phillips. He was a colorful gentleman from Texas and a delight to work for at FIRST Fleet."

"Admiral Nimitz was a land-locked Texan when he started as well. Is there something about the Lone Star State that makes for great Naval leaders?"

"Admiral Phillips was as authentic as they come. He wore a Silver Star for valor and had, at various times, commanded a submarine, destroyer and a cruiser. I think he was Annapolis Class of '17, and had World War I service on a cruiser. He was assigned to intelligence duty in the Panama Canal Zone, so he knew a little bit about what we do, and in the second war he was in command during the Gilbert and Marshall Islands campaigns, the first battle of the Philippine Sea and the later strikes again Guam and Saipan. And you might be right about Texas. I have worked often with the nice folks at the National Museum of the Pacific War, which is located in Fredericksburg, the boyhood home of Admiral Nimitz. Not many of us left who actually knew him."

"I actually had an interaction with them when I was at THIRD Fleet the first time," I said, taking a pallet-cleaning sip of wine. "They were looking for a piece of the *Arizona* to display in Texas. One of the Pearl Harbor secrets is that the wreckage cut off the ships after the attack was taken to the Waipio Peninsula and stacked in pieces according to what ship it came off of in case it was needed later. There apparently had been periodic attempts to purchase the pile of rusting steel that had been Arizona's superstructure. I suppose the buyer wanted to make souvenirs out of the steel—key chains, maybe."

"Clearly an inappropriate use of parts of a war memorial," said Mac firmly.

"For sure. But I wound up with the chart of where the pile of Arizona steel was, and the museum sent someone out to look at it for something that would be immediately recognizable as being part of a ship. They found a hatch and frame that would do nicely, once cleaned up and painted haze-gray."

"I don't recall seeing something like that the last time I was down in Fredericksburg."

"No, I don't imagine you did. Someone stole it from the warehouse while it was in transit to Texas. So maybe there are some keychains out there we don't know about."

"Sol Phillips would never have put up with it," said Mac. "Like I said, he was a character. Destroyer-man in the Big War, much in the manner of 31-knot Arliegh Burke and his Little Beavers. Wyman Packard knew him better than I did, since he had been at FIRST Fleet for a couple years when I showed up. He was given a fourth star when he retired and went home to Texas. But my favorite story about him is what shows the kind of guy he was—real Old Navy. When he arrived to take command, Wyman scheduled him to be indoctrinated for communications for the first time."

"Imagine that—a three star admiral and combat hero and no

Special Intelligence background? That is how commanders make uninformed decisions."

Mac nodded in agreement. "That is why it is important for the Admirals to know about all the arrows in their warfare quivers. But since Admiral Phillips had never had access to Special Intelligence before, he had to have a background investigation. The Navy insisted."

"I hate those things," I said with a sigh, thinking back to the time I got in trouble for falling asleep during a lie detector exam. Not to mention the urine tests and all the rest of that nonsense. Like it stopped Snowden or Manning from walking off with the whole database."

"It is what it is," Mac said. "Has been since the beginning. Anyway, since a background investigation of the Fleet Commander was required by regulation, the machinery began to grind on. Forms were filled out and submitted and the Navy investigators began their work."

"I am actually kind of looking forward to not having a clearance," I said. "The five-year general updates, the five-year polygraph cycle, and the two-year Top Secret in-status limit. I am tired of my own government following me around."

"You have to be careful with the Crown Jewels," said Mac. "But remember, the very existence of COMINT was not declassified until the early 1970s, and it was a very big deal when I was at FIRST Fleet."

"I understand. So how did the Admiral take it?"

"He told a story that he heard back from a buddy in Texas while the investigation was in progress. What he heard was this: "Sol, you ought to know the 'Feds' were here asking all kinds of questions about you. And I didn't know what you were up to so I told them everything I could to steer them wrong. I hope they never find you."

"Did he get his clearance?" I asked.

"Eventually, he did, and Wyman read him into the program, and he continued his command tour. He used to have a sign in his office that read: "They who go around in circles shall be known as Big Wheels.""

"Our pal Jake had an inflatable cow head mounted on the wall in his office as the Director of DIA."

"I would expect nothing less," and he took a sip of tomato juice. "Did I mention the best restaurant in Coronado, the Mexican Village?"

"It was still open down by where the ferry landing was. I heard that before the bridge was opened, people would get the cars in line to board the next one to go to the San Diego side and have a margarita or three while they waited."

"We called it the Mex-Pac, in honor of where the fleet units were deploying. Great place and great food."

"It was. So much is gone now." I thought about the other little bar just up Orange Avenue from Mex-Pac, the one that opened at 0600 to catch the Chiefs trade before morning quarters. "So,what does a fleet commander concentrate on when he is home-ported in San Diego?" I asked. "I know what we were interested in when I was there, and that was mostly Soviet ballistic missile submarines."

"Admiral Phillips concern was primarily Southeast Asia. He had just returned from there, as I mentioned, and I think he saw that there would be a wider conflict presently. He wanted to be fully prepared if he was ordered to deploy again and set up a joint staff to manage operations in Vietnam."

"What sources of information did you use to keep the Admiral up to speed?"

"All sources, attaché reports to COMINT. Communist expansion in the region was a concern, and a primary one for the Admiral. But we also kept an eye on the Soviets and particularly the Soviet Pacific Fleet."

"It became an all-consuming effort," I said, finishing my glass of

Happy Hour white. "But working for a commander who understands what is going on and has a vision to deal with it is pretty special."

"You bet. We could level with Sol Phillips and tell him what we really thought. Sometimes you have to do that. Admiral Phillips was a great man to work for, and we were sorry when he retired."

"I bet. Is that going to get us to Vietnam?" I said hopefully.

Mac shook his head side to side. "No, that is going to mean a trip to Arlington Hall Station and Fort George Gordon Meade first."

I picked up my pen again and waved at Big Jim down the bar for reinforcements.

DET ALPHA

● ● ●

SABRINA, SPIRITUALIST AND VENDOR OF SPIRITS-BOTH
LIQUID AND OTHER-BEHIND THE WILLOW BAR.

S abrina was on duty behind the Willow bar. She is an intensely spiritual mixologist, and you can always get your palm read or get an account of the full karmic impact of whatever is happening on our side of the bar from her. Saturn was apparently in the 11th House, and from what I could gather, that was important because of that planet's influence "on the right ear, the spleen, the bladder, the phlegm, and the bones." I knew I was going to be needing all of them, and I thanked her for the information.

I enjoyed the first sip of wine for the day as I waited for my weekly interview, eager to get on with the story of the late 1950s, and hoping to arrive in Vietnam before I ran out of the ability to safely consume the Happy Hour White. RADM Mac Showers arrived presently as we discussed what Saturn might mean for the rest of the week. The Admiral was looking dapper, and he almost asked for an Anchor Steam beer because he was feeling so good. I got organized next to him—white wine where I could reach it, plenty of cocktail napkins on which to take notes and a working pen in my hand. As with nearly all of these discussions, I had hoped to start one place and we wound up somewhere completely different.

"We talked about FIRST Fleet the last couple outings," he said. "But there was another aspect of readiness and training that I developed while I was on the staff. That is the one I would like to talk about in some detail. In talking to the various elements of the fleet about readiness for deployment to the Seventh Fleet and what they would be requiring once they got to WESTPAC, we realized that the aviators were drastically short of adequate target materials in order to carry out the nuclear strikes for which they were responsible, and for which they were doing a considerable amount of training. They simply didn't have the materials for training purposes, or, once they were deployed, they didn't have good target materials to carry out long-range overland strikes into the interior of China or the Soviet Union. So, this became one of my missions, to find out what they needed and to take necessary steps or to have developed better current updated target materials. With the assistance of my aviator assistant N-2, we took it upon ourselves to visit the bases and training facilities of all of the aviation groups that had nuclear strike responsibilities where they were based on the West Coast and where they were carrying out their training activities. Working with them and through AIRPAC, we conducted an extensive series of interviews of the pilots themselves, as well as the squadron and air group officers, to find out exactly what they needed in an ideal situation to carry out their strikes, and we then used this information to devise a target folder that would include all the materials they needed."

"That can be a challenge. I had an A-7 pilot on the *Forrestal* Med Cruise in '90 who didn't like the satellite image we were able to get of one of his targets. He said he would be approaching from the other side of the power plant he was assigned to take out, and asked if I could get him a picture of the target from the other direction. I realized I couldn't do it—not while we were on cruise, anyway. It would have meant getting a target nomination up through SIXTH Fleet, and into the scramble for priority with the Imagery Committee back in DC. Who did you go to for support?"

"We worked with a number of people. We worked with our fleet Weather-Guesser, who was actually assigned in the plans division of the staff. We worked with the other officers in the Plans Division, who included an Air Force officer as well as an Army and Marine officer. Through the Air Force officer I even made a trip to SAC to find out generally the types of materials they put in o their target folders for their strategic bombers. This gave us a lot of ideas about photography, aim-points, check-points, turn-points, all the things you needed to get an aircraft from a carrier 'feet dry' and into his target. Of course, the Navy delivery tactics and platforms were substantially different than the Air Force was using with their high-level, long-range strategic bombers. So, we had to modify our target folders accordingly. We did that, and we established in detail the materials the pilots wanted in order to carry out their missions. Having done that, we *then* encountered the problem of how to produce these target folders. We had the target list. We knew what the targets were, and where they were. We just needed to get the images, charts, and checklists to the pilots who had to fly the missions."

AD-J SKYRAIDER IN FLIGHT, ONE OF THE SLOWEST AVIATION DELIVERY PLATFORMS FOR NUCLEAR WEAPONS. DAD TOLD ME THAT STRAIGHT AND LEVEL, YE COULD OUTRUN A P-51 MUSTANG, AND CARRY MORE ORDNANCE THAT A FLYING FORTRESS. IF I WAS GOING TO LIGHT UP A TARGET IN THE SOVIET FAR EAST MILITARY DISTRICT, I THINK I WOULD LIKE TO HAVE A JET UNDER ME.

"So, who were the people who designated the targets? Was that done at the national level or with the Strategic Air Command?"

"Oh yes—it was all part of the Single Integrated Operational Plan."

"I knew the SIOP. The people in Omaha were always running drills during their work-day, which as a pain in the butt when you are in the Far East someplace. They would go all night sometimes when we had the Emergency Action Message watch on the ship."

"The detailed information was held at PACFLT and at FICPAC intelligence-wise, not by us. In fact, it was with only with some difficulty that we were able to penetrate the bureaucracy surrounding the list."

"Did you succeed?"

ADMIRAL FELIX STUMP

"To some extent, we did. But I was extremely fortunate in having Captain Rufus Taylor as the N-2 at PACFLT, and he was extremely cooperative and helpful in opening the doors for me. The Plans Officer at PACFLT at that time was then-Captain Grant Sharp, who later became Commander-in-Chief at PACFLT and then the theater commander at PACOM."

"Some people just can't bear to leave the Islands!"

"He was the Plans Officer at PACFLT at the time and Admiral Felix Stump was the Commander-in-Chief. We had to convince each one of these officers through a series of briefings about what we were doing, what we needed, and what type of information to which we had to have access. With their cooperation and a lot of pushing by Captain Taylor, and extensive briefings, we did, in fact, push forward to gain access to the necessary information simply to be able to devise target folders. But then the physical production on these targets became the next problem. First Fleet did not have the manpower nor the facilities, nor the information in San Diego, to do this job."

"Who did? The Fleet Intelligence Center at Pearl Harbor?"

"Just so. The information was held in Hawaii; the raw material was largely held at FICPAC, where there were a large number of people working over photography and various other things that were needed in these folders. They had the photo-interpretation facilities and all of the equipment, none of which we had in San Diego. But despite the fact that the information and facilities existed in Hawaii but not in San Diego, CINCPACFLT still said it was a FIRST Fleet responsibility, because it was a readiness responsibility, and they said, "You proceed to do it, plan it, and do it, and we'll support you."

"Sounds typical. So, who did?"

"In the end, the solution was to create a new organization in Hawaii that was co-located with FICPAC, which became known as FIRST Fleet Detachment Alpha, or just "Det A" for short. That was created and billets were procured from thither and yon within the

Pacific Fleet and the air organization and FICPAC and PACFLT and so forth."

"Birth of a new command billet. Who got the rose pinned on him to take charge?" "An aviator named Jack Fitzgerald. Jack was a gung-ho aviator who had been in our

Readiness Division on the FIRST Fleet staff for more than a year. He and I had become

close friends. We had taken several trips together. He had assisted us in devising the content of these target folders as we developed them. When we reached the point of creating First Fleet DET Alfa, he was eagerly anxious to head it. It was a natural; he was an aviator, and he knew the problem. He was interested and competent, and he wanted to move to Hawaii. So we did; he became the officer-in-charge.

He moved to Hawaii, and he was the only billet that we gave up to form the Detachment. All the other billets, and I think we got twenty or thirty people who made up this detachment, came from within PACFLT resources. They used FICPAC's materials, photos, graphic materials, and used FICPAC's equipment, photo-interpretation equipment, drafting tables, and so forth."

"And we literally, physically, produced target folders, one by one, by one for the designated targets, including the materials that we had identified that needed to be provided to the aviators so that they could study their en route targeting, their check points, their delivery points. all of which was an art form that had been developed for nuclear bomb delivery in those days. Of course, it was different for different types of aircraft. We had the A-3s that were large twin-engine jet bombers, and we had the ADs, which were a single-engine, propeller-driven plane that would fly long legs at low altitude and deliver their bombs by a pop-up mode; an 'over-the-shoulder' delivery it was called in those days. So we had to devise slightly different target folders for different types of aircraft."

A-3 SKYWARRIOR

"Those were the days of the Navy's first adventure with cruise missiles, the subs that carried the Regulus missiles. Did you also have to produce targeting material for them?"

Mac sighed, remembering. "Yes. We tried, but we didn't really ever get into that because they didn't need anywhere near the amount of information, and the type of information that a live pilot flying an airplane had to have. I don't recall where their targets were designated. We didn't do that; we didn't designate the Regulus targets. But I believe that they were provided with target materials by whatever authority was designated their targets. That was never a very extensive program, and they didn't have very many targets since they didn't have very many weapons."

"Well. There were not that many Soviet targets in the Far East beyond Vladivostok and the cluster of submarine targets on the Kamchatka Peninsula."

"True. As I recall, the ones we were generally concerned with were operating near Petropavlovsk, and in the northern reaches of the Soviet Union. The ones around Vlad in the Sea of Japan were air targets."

USS *GRAYBACK* (SS-574) UNDERWAY ON REGULUS MISSILE PATROL.

"We mostly worried about Petr, where their front-line boomers and attack subs were based.

But it sounds like this was the birth of the target folder era for the carrier pilots."

"Yep. The target folder program and the creation and management of First Fleet DET Alpha was really a major development and became a leading steady-state activity in the intelligence division during the two years that I was there."

"I know generations of intelligence officers who have spent time helping to build those folders, Admiral. It was the main part of any squadron or Air Wing job."

"It was a stimulating tour. It was a job in which, for the first time, I really felt like I was given a free hand and very good support in developing some innovative things—mainly these target folders, which had not existed previously. It was unbelievable to me when I realized what the Navy pilots were expected to do and how little they had to do it with. When we used to sit down with pilots who were flying these planes, and told them that we would provide them target materials so that they could do their job, they were just astounded. This was the first time anybody had come to them and said, "I'll help you."

"Better than telling them to 'fly west 'til you see a naval base,' I suppose."

"That is for sure. It was also heartening to know that we were doing something that somebody really appreciated. Of course, the program was still going on in high gear when I left, and I don't what the eventual outcome of it might have been. Except that I *do* know that a few years later when I got back out to PACFLT as the N-2, I discovered that DET Alpha had been disestablished, and FICPAC had, in fact, taken over the target folder preparation, and they were doing it, which they really should have done originally."

"You almost had the tail wagging the dog, didn't you?," I said, waving to Sabrina down the bar for emergency replenishment of the Happy Hour white and some positive psychic energy. Talk about nukes always leaves me a bit drained.

The Admiral gave me a thin smile of satisfaction. "Well, I don't know what FICPAC was doing at the time that was so all important, but 'Rufe' Taylor said that he simply couldn't take on this mission as an additional function and that he would help us create Det. Alfa if we would get the people and staff it. Well, we did. He helped us get the billets, he provided the spaces, and we got photo interpreters from various fleet units. We found them were scattered around the

Pacific Fleet and many were not even working in rate looking at imagery. So we were able to beg, borrow, and steal photo-interpreter talent and got them assigned to our detachment."

How much support did you get from the Office of Naval Intelligence back in Washington? That seems like it would be a national mission."

"Very little. It wasn't their concern; it was really a Pacific Fleet show. I worked with an organization in San Diego that..." Mac looked up, a bit startled that his memory had failed him. "I can't remember the acronym for it, but it had control of all personnel assigned to all ships and units in the Pacific Fleet, and if I said I wanted to find out where all the

Photo Interpreters are, they could give me a computer printout of a hundred or so photo-interpreters who were scattered through-out the fleet. It was a success story. I never did know if the people on the Atlantic side ever did anything like this. We were pretty autonomous."

"OK," I said, putting down my pen to concentrate on the wine for a moment. "Two great Fleets, Atlantic and Pacific, separated by a common Navy. So that about covers your time in San Diego. What was next?"

"Back to Washington. And back to Special Intelligence and Y-1, finally. After having left Y-1 back in 1950 to go to London, I now had orders to come back and *head* Y-1. I was delighted to come back because I really felt like I was coming home after having been around the world in Europe and detoured in the Intelligence School then out to San Diego. It was with great delight that in 1957

I headed back to Washington. As a matter of fact, I had the family already back there because the house that we owned at that time had been rented to a Navy Commander and his family, and they left in December on assignment and we decided to let the house stand vacant until we got back."

"Ah, the joys of the military move!"

Mac laughed and reached for his Virgin Mary.

VIGILANTES

• • •

AN RVAH-12 RA-5C BEGINNING A RECONNAISSANCE
RUN OVER VIETNAM, 1967. PHOTO: USN

M ac was waiting patiently at Willow, halfway down the bar that Old Jim anchored at the Amen Corner. I was a little flustered. A subcontractor to us on a counterintelligence contract down at the Russell Knox Building at Quantico had some people working who got bored and staged an "unattended package" prank on a co-worker. When it was discovered, the building was evacuated—the whole thing, not just the people working on our little task order.

The end-of-day calls revealed that the Government was expecting to be reimbursed for the lost time of everyone in the building, and that I was generally responsible for all of it. This was going to be a mess for a couple weeks until we could unscramble everything.

Sometimes I hate this business. I called Bronco to express my outrage and vent on the situation.

Bronco is one of my earliest Fleet buddies. He was one of *Midway's* Air Wing FIVE Landing Signals Officers, and he flew with our version of the Vigilantes to stay current. Later, he had his own Squadron and Deep-Draft command tours before he retired from the Navy as XO of the Saratoga. A gifted fighter pilot and exceptional officer, Big Navy retired the last of the conventionally-powered carriers, and without the nuclear power certificate, he got aced out of a chance to command for command of one of the Nimitz-class boats. No question in my mind, he would have made Flag in a walk.

In the world of retirement, he wound up like I did—as a Parkway Patriot, working on bid-and-proposal tasks and wearing a goofy bow tie like me. It was a fashion statement that shouted out: "GFY." When he has business in the Willow neighborhood, he will stop by for a beer. When I finished venting, I told him Mac would be at Willow that evening, he said he would be there. Which actually is what made this tale from Mac's memory happen the way it did.

After I dodged death to cross Fairfax Drive against the rush hour traffic, the cool darkness of Willow was a relief. I stopped to exchange greetings with Jim, who was content to sip his long-necked Bud and listen to old jazz on his MP3 player. I promised we would catch up after my chat with Mac. I walked down the bar and pulled out one of the stool and sat down after shaking his had. I started to gather the critical materials I needed to conduct the interview-a pen from my jacket pocket, a stack of cocktail napkins, and my smartphone in case we needed information on the fly. I don't know how people did histories before smartphones. Honestly.

Mac was dressed neatly in a sports jacket and tie. I told him that Bronco might be joining us. Mac brightened when he heard that he was a Vietnam-era fighter pilot. "We will talk about airplanes, then. We were talking about FIRST Fleet the other day. We were in the process of trying to make aerial reconnaissance reflect the changing nature of the Cold War.

Advanced surface-to-air missiles were changing the equation for strike targeting purposes, and recce to assess the Bomb Damage Assessment. I was thinking about the RA-5C Vigilante program after we talked."

I had my pen out and wrote the subject of the meeting when Bronco appeared at the door and peered in, looking for us. I waved, and he came down and took the stool on the other side of Mac. I made the introductions and he asked for a craft beer from Liz-S behind the bar. I took a sip of Happy Hour White and said: Bronco, we were talking about the RA-5C. We did not have them on Midway. Too big a jet for too small a flight deck. We had to use the Marine RF-4 Phantoms with camera capability to do our post-strike and recce missions. The RF-8 Crusaders were out of the Fleet by then."

"I can give you some context on that," said Mac. "The RA-5C program was derived from the attack version of the jet, which was intended to be a supersonic deep penetration aircraft that would keep Navy in the nuclear strike business and not let the Zoomies claim that all nuclear warfare could be done by the Air Force."

"I heard that didn't work out that well. For survivability, the strikes were supposed to be a Mach speed, and the weapon was deployed out the after end of the jet. Problem was that at speed the bomb would follow right along in the vortex."

"You are correct on that, Vic. The Vigilante was built by North American in Columbus, Ohio. The program had its beginnings when I was in Y-1 and when Rufe Taylor, as a Captain, was head of the Intelligence Division of ONI. All of us were involved in the planning and structuring of the RA-5C program."

"You were providing the requirements for recce, right, like for the satellites?"

Mac nodded. "We even made trips out to Columbus where it was being built, to work with the contractor. Of course, it was both a strike aircraft as well as a reconnaissance aircraft. We were concerned with the reconnaissance configuration; it was the RA-5C and had a canoe-like pod attached to the underbelly that would hold the sensors. The whole program, of course, was to have a platform that could be carried on a carrier, and launched to conduct its collection mission in a tactical environment and return to the carrier with the yards and yards of film from the canoe. It would be read out on light-tables right there in the Integrated Operational Intelligence Center. Not exactly near-real-time, but that was the best we could do for the crews that had to carry out the strike missions. Completely organic, and on the carrier. We did not have to wait for the Fleet Intelligence Center to turn it around to us. And remember, there was no imagery except hard copy."

"I remember the Special Access package that FICPAC or the 548th Reconnaissance and Targeting Grroup (RTG) at Hickham would provide. It was the only satellite imagery we had for the cruise, unless something came up important enough to fly it out. We were literally cut off while we were at sea for most things."

"Different world now, I imagine," said Mac. "We never had anything fancy like that, and that was just the way it was." He took a sip of his Virgin Mary. The Oncologists had him on the wagon again. I took a sip of Happy Hour white, swirling it on my tongue while thinking what version of the riot act I was going to read to the subcontractors who had played the prank. And made a note to talk to the lawyers at Corporate. "When I was in PACFLT after 1962, the RA-5Cs began deploying. I want to you to note this," he gestured at my napkin, "because, on one occasion when I was N-2, my friend, Gene Fubini, who was then Director, Defense Research and Engineering (DDR&E) in the Pentagon, came out to Hawaii on a trip."

"A boondoggle," I said. "Seems we had a lot of Washington people come to Hawaii for critical meetings in January and February."

Mac smiled. "Gene visited our headquarters for briefings, and, as I was seeing him off from the quarterdeck of the head-quarters, the USS *Constellation* (CV-64), was moored right down at the foot of the Makalapa Hill at the Carrier Pier. Connie had an RA-5C sitting on its flight deck, clearly visible. Gene was a great advocate and supporter of the RA-5C program, and he looked down there, saw that aircraft and he said, "Mac, now that we have that aircraft deploying, we'll be able to decommission the VQ squadrons."

And I said, "Dr. Fubini, that will never, never happen." He was convinced it *would* happen though, and *should* happen, and I was convinced it wouldn't because the two programs were so different in mission and capabilities. His concern as DDR&E was that a lot of research and development money had been put into the Vigilante system; it was new; it was a tactical support system; and I think he felt genuinely there wouldn't be a need for VQ collection after the RA-5C program was deployed."

EP-3 DISPLAYING RADOME AND SPECIAL ANTENNAS. PHOTO: USN

"He probably would have liked unmanned drones, too," I said.

Bronco had been listening intently, and cleared his throat. He spoke quietly. "In 1972, I escorted a Viggie on a recon mission from Ranger-Maru over The North. The Viggie was so clean—little drag—that it could easily go the "speed of heat." The pre-flight brief with the Viggie driver was said that if we got shot at, he would only use basic engine (no afterburner) and I would be able to use burner to keep up. Didn't turn out that way. We were poking along at about 450 knots when we did get lit up by a fire control radar."

I stopped writing, thinking about what it had been like having nothing but some aluminum between your butt and an SA-2

Guideline. I put down my pen as Bronco continued, looking across the bar.

"Well, the *Viggie* immediately lights the his burners. I light mine and try to keep up, but it was futile. After all, I have four pylons, a centerline tank, two Sparrows, and four Sidewinders out "in the breeze." After a couple of minutes, the Viggie was a supersonic dot on the horizon. I said: "screw this" and headed for the coast—and started asking for a tanker."

I felt like I wanted to shiver. Mac smiled, remembering Naval Aviation stories from long ago.

"By the way," Bronco continued, "the Viggie was also a very intimidating bird coming aboard—for a LSO like me under training. It was so big, so fast, and—would you believe—so fragile that the LSO and the driver had very little margin for error. I am still surprised that the CAG LSO ever let me wave them at night."

Bronco had once invited me out to the LSO platform to observe a night recovery. There is something very intense about watching jets make that approach and plant themselves with such precision in the landing area. One nugget was having some problems with line-up and glide slope, and I leaned forward in fascination as the dance as done to bring the jet aboard. "Wow, I said, that was close!" I looked around to see the LSO team's reaction. There was no one there but me. They had all jumped to the netting below the platform because they knew what disaster had been so close. I respect those who know the potential consequences of what they are doing and do it anyway.

"Well, I'll be the voice of history tell the rest of the story." Said

Ma. . The RA-SC program worked effectively for a few years, the VQs continued, the RA-SCs are now museum pieces, and the VQs still continue to fly."

"They are going to replace them with a jet in a few years, but some of the EP-3 *Orions* still flying were manufactured well before the Vigilante."

"That's right. As good an idea as the RA-5C and the IOIC program was, in the totality of naval intelligence, I think, it was just a passing fancy."

"It had passed by the time I got to Midway in 1978. We worked in what was called the Carrier (CV) Intelligence Center—IC. The nickname was "CIVIC." The Vietnam-era pilots still knew the IOIC by the nickname "101 Clowns.""

"The basic RA-5C airframe had problems that made it difficult to use as a bomber, and it took up so much room and was just so limited in its applications that it couldn't continue to compete for space on the carriers. Plus, the things that it did, or was designed to do, were taken over by satellites so effectively, and in even better time--closer to real time--than what the RA-5C could do that it became superfluous."

"I knew we all got confused. My F-4 Phantom squadron dated back to World War 2, and our call sign was the "*Vigilantes*."" I looked at Liz-S behind the bar hopefully. "We got confused with the recon guys all the time when the Cruise Boats would come out from the West Coast."

"It can be a confusing business," said Mac. "I am just glad to be here to put things straight." Bronco smiled and finished off his beer.

WHY ONE

• • •

Peter was behind the bar, which means that this undated interview happened sometime before he either got fired or decamped downtown on his own volition. There is no one who could pour a more infuriating martini—the surface tension alone held the precious liquid in the glass, making it almost impossible to take the first sip with any grace. Normally I would use one of the small cocktail straws to draw down the level a bit and make in manageable, even if I had to hunch over the bar like Quasimoto to accomplish it.

I miss Peter and whoever his little assistant was—I remember they dressed up as Batman and Robin on one of Tracey O'Grady's spectacular Halloween parties. He was the first of the long-time bartenders to depart and make our little circle of barflies smaller. So, given the bar staff, I have to guess that this was a conversation in 2008 or 2009, back when Mac Showers and I were just getting rolling on the project.

HEADQUARTERS BUILDING AT ARLINGTON HALL STATION, ARLINGTON, VA.

He talked a lot about Arlington Hall Station, and it resonated with me, since the station is just across the street from the building where I live, and the cemetery just up the road contains the graves of many Japanese-Americans, which Mac explained was not unexpected, since the neighborhood was a nest of spies for years since the Army took over the former girl's school to make it their number one code-breaking installation, the bookend to Navy's occupation of the academy for young women on Nebraska Avenue.

I learned my lesson on Peter's martinis, which were as potent as they were full, and was careful to stipulate the Happy Hour white wine, which in this case was a nice Flora Springs Napa Valley Chardonnay, 2004. At five bucks a glass, it was a good deal and kept us all coming back. Well, that might not have been the reason, since Jon-Without-an-H drank vodka and soda and Old Jim would not tolerate anything but Budweiser in a long-neck brown bottle. No accounting for taste.

Confirming that it was a Napa wine by the color of the stains on the napkins on which I had taken notes. I honestly do not know how we got to taking about the Navy code-breakers who had moved into Arlington Hall after the war. The new and chilling relation with the Soviet Union brought the National Security Act of 1948, and a new consolidated organization to handle the COMINT missions of the military departments. The new organization was dubbed the Armed Forces Security Agency (AFSA), and was stood up in DoD on May 20, 1949."

"I have visited the Key West White House where Mr. Truman worked on the Act. I think I would have just stayed down there. Was the process as confusing as I think it was?"

"Short answer is 'Yes,'" Mac said. "In theory, the AFSA was to direct the communications intelligence and electronic intelligence activities of the military service signals intelligence units, which now included the brand new Air Force. In practice, the AFSA had little power, its functions being defined in terms of activities not performed by the service components. The answer, it appeared, was the creation of the National Security Agency, which sprang full-blown from the forehead of Walter Bedell Smith in a December 1951 memo to James B. Lay, Executive Secretary of the National Security

378 ••• *Vic Socotra*

Council. The memo observed that "control over, and coordination of, the collection and processing of Communications Intelligence had proved ineffective" and recommended a survey of communications intelligence activities. The proposal was approved on December 13, 1951, and the study authorized on December 28, 1951."

"Aha!" I said triumphantly. "Our first Blue Ribbon Panel on intelligence failures! Weren't some of those Navy Radio Intelligence guys who screwed Joe Rochefort part of the leadership?"

"Yes. They certainly were. And that certainly was not the last of those things in my career," Mac said with an ironic smile. "The report was completed by June, 1952. We called it the "Brownell Committee Report," after the chairman. The thing was a survey of the history of U.S. communications intelligence activities. It came to the conclusion that there was a need for a much greater degree of coordination and direction at the national level."

"Time for a name change, which is often the only way to move the deckchairs around on a sinking ship. The role of NSA was clearly going to extend far beyond the armed forces, and the facilities at Arlington Hall were not going to be adequate to the scope of the new mission." At least that is what the histories say. I preferred just hearing it from a guy who was there.

We may have gotten to this since I had been relating a story that Louie, the former Chief of Staff at DIA, had told me about being a young analyst at Arlington Hall. When DIA was just starting and housed on the Arlington Hall campus, he worked in the Soviet Shop with a Navy Commander who had the irritating habit of removing his shoes when he came in the office, only putting them back on when the work day was done. When it came time for his to execute his Permanent Change of Station orders, Louie got even by nailing the empty shoes to the wooden floor of the 'temporary' office building. Mac was mildly amused, and mentioned that the windows on the Secure Compartmented Information Facility were drafty—the only time he had that problem with a SCIF. I asked him about his experiences at Arlington Hall.

"Well, I worked at Y-1 I worked in Y-1 between '46 and '48. Admiral Frost was the DNI when I came back to Washington from FIRST Fleet, and I had worked for him, of course, when he was a Captain. He was originally Y-1 in 1946, and I saw him frequently until I went to London. Admiral Frost, in the mean time, had gone off to do fleet duty or whatever. It was non-intelligence assignment, and then he returned after making admiral as the DNI. 1 am not sure of the year. But he was the DNI when I returned to Washington in the spring of 1957 to take over Y-1. I found Y-1 much as I had left it, except enlarged. There were more people, different people, although some were the same.

"Maury Hellner and Ed Nielson were still there, Dr. Ed Haff was there. All of them had been hired into Y-1 as civilians; those three were civilians when I was there the first time, and they were *still* there. Maury stayed there during most of my tour, and Ed Nielson continued for a little while. Ed Haff had, I think moved over into the Pentagon into "F" branch just after I got there. But all those three people, who became high-level analysts and well-recognized in ONI were first hired in naval intelligence service in Y-1 and cut their teeth doing Y-1 analysis. They did a creditable job all the time.

The active duty officers assigned there were, of course, rotating, so the fact that I had been gone from Y-1 for seven years meant only the civilian cadre remained. Well, except for LCDR Barbara Conard—her married name was 'Moore,' if you want to check her out. But she, too, had other assignments while I was gone before returning. She was one of the original Y-1 group that we had at Nebraska Avenue in 1948."

"I knew one of the women who worked there in the war. She wouldn't talk about it, and freaked out that I knew about it, even though her work had been declassified decades before. Was she an intelligence officer by designation?"

"No. Barbara was a WAVE. And you don't use that term anymore, I don't believe. You call them "women officers.""

"They changed that to General Unrestricted Line—pronounced "GURL," which seemed mildly sexist to me when I was in the personnel business, and trivialized the contributions of the WAVES. But now everyone can be a warfare specialist, even the Spooks. They even have a little badge to wear that says so."

"I started as a Deck Officer, so I understand. But Barb was a Wave. She did not have a Special Duty-Intelligence designator. She had been on an assignment in New York and someplace else and then she came back to Y-1. I think she had been there about a year when I arrived back in 1957, and she stayed another year and then she received another assignment. But, other than that, a lot of the people were new. We had two or three Ensigns who were Naval Academy graduates but because of physical deficiencies, mainly eyesight, had been commissioned as intel and then had gone through Intelligence

School. That program started when I was at the Intelligence School, and it had continued with three or four such officers a year. Some of them were on duty in Y-1, and some of them became career intel officers. That includes Admiral Bob Schmitt, who first worked in Y-1 as an Ensign, as did future DNI RADM Torn Brooks. And we moved to Fort Meade. The organization was larger, but the organizational structure that we had originally created still existed. There were some additional mission areas and simply more people assigned to do the increased volume of work. The products being put out were increased because of increased personnel, and the growing volume of raw material being provided."

"What accounted for that, Sir? New collection systems coming online?"

GRAB Satellite, forerunner of orbital ELINT collection systems.

"Yes. There were land-based sensors that provided comprehensive HF/DF information and information on Soviet radar emissions and communications. Navy was the first of the Services to go to space with the 1960 launch of the Galactic Radiation and Background (GRAB) satellite, the cover name for the first of the ELINT collectors. POPPY was the follow-on system, which might have still been on orbit when you joined up. That was the mainstay of the U.S. Navy's orbital ELINT collection capability for almost 15 years. It started out as a general search system whose mission was to map out the locations and capabilities of Soviet radar systems, but gained capability and data volume over time as new electronic systems were added to each launch."

"It became a sea of data and now they are talking about it all floating around in a cloud. I am glad I no longer have to deal with it," I said, sipping some of the Flora Springs.

NSA's Operations Building #1, Fort Meade, MD

"So it was not hard to fit right back into Y-1 when I returned, and we lived in North Arlington, so the commute was a breeze. The first issue I was confronted with after reporting was the fact that we were soon going to move to Fort Meade. I had no prior knowledge that NSA was building a headquarters building at Fort Meade and all that went with that. But when I arrived there in the spring of 1957, the construction at Fort Meade was well along. The head-quarters building was due to be occupied later in the year. Many elements of NSA had already moved. When they created the NSA compound at Fort Meade, the first thing they constructed was the barracks for the Marine guards. These were three or four reinforced concrete, two-story buildings near the NSA headquarters building. Several elements of NSA were temporarily housed in these barracks buildings until the headquarters was completed. So, at the very beginning of my tour that Spring, we had to make liaison visits to some of the NSA elements at Fort Meade, and we went out on a few occasions and worked with the NSA people in these barracks build-ings, which was rather disorganized and temporary, but nevertheless NSA was already operating on a split basis. So, we continued our planning for the move through the year, and, in December of 1957, we actually made our physical move. All the rest of NSA moved, and Y-1 moved along with it, in December 1957 and everybody took up full residence at the new headquarters building at Fort Meade at the same time. That opened a new era of activity with Y-1, which in the mean time had been re-designated the "Navy Field Operational Intelligence Office." I always have to stop and think because of the difficulty of that title and the artificiality of it. I still to this day refer to the activity as Y-1, and I believe I always will. It's an easier thing to say."

"I totally agree about NFOIO's name. But we had to acknowl-edge and use it, since they were the Gods of OPINTEL when I was out in the Pacific. What they said about the Soviet subs and merchant shipping was considered authoritative, and us Spooks out in the hustings were just the Junior Varsity."

"That was the nature of the Ocean Surveillance Information System."

"I am proud that I got to be part of the OSIS system," I said, "We used to say that working at one of the Centers or Facilities was like trying to write a newspaper on the floor of the New York Stock Exchange."

"Dave Rosenburg wrote a book about it called 'The Admiral's Advantage,' said Mac. "He talked to me extensively on the origins and function of the system. We had the intelligence nodes of the system around the world—London, Rota, Pearl and Kami Seya. Sitting on top was Y1. You should be proud that you got to see one of the finest OPINTEL systems ever devised. But as to the NFOIO name, it was all Beltway politics and centers of gravity. At that time, Fort Meade was outside of what was defined as the metropolitan area, Washington, D.C. Fort Meade was geographically within the area of the Severn River Naval Command. I don't know whether the Severn River Naval Command, as an administrative organization, still exists.

"Most of those commands have long gone away," I said, wondering if there was a good time to step out and grab a smoke.

"I haven't heard of the Severn River Command in years, but it was real, and led by the Superintendent of the Naval Academy. He was double-hatted to run the Naval Academy and the Severn River Naval Command. We fell within his geographical limits, so that was the reason that the activity was designated the field office of ONI located outside the Washington metropolitan area and why I was given additional duty orders to report to the Commandant of the Severn River Naval Command, who technically was my military superior."

"And with whom you had very little business?"

"Almost nothing. I did go down to Annapolis to show we could play nice. I took Commander Ed Cummings with me, he was Y-lE, my production manager, and he was an Academy graduate. I'd never been to the Naval Academy before. I didn't know my way around, so I figured I could use a ring-knocker to keep me on course."

"Just to stay calibrated on this napkin, where were you in your career? Did you have your twenty years at that point?"

"Right on twenty in that tour, and just made Captain, so I had no thought about retiring. I was having too much fun. So, I took Ed with me to make my call on the Super intendent and to do all the necessary protocol niceties. And we did that; we had a nice discussion with him. We couldn't tell him very much, because he wasn't indoctrinated for our business, but he wasn't overly inquisitive, so I'm sure he had many things of his own concern, and we were not very important to him. But he said he would designate one of his staff officers to be our point of contact for whatever business we might have. He designated Captain "Lucky" Fluckey who had already had some intelligence duty. I think that's why he designated him.

"That would be 'Lucky' Fluckey? The submariner who landed a sabotage party ashore in Japan to blow up a train during WWII?"

"And later Director of Naval intelligence. I knew him slightly at that time, and he was on duty in the Academy's Electrical

Engineering Department. So we continued our call and went over to see Captain Fluckey, and he was very receptive to us, very helpful, and said he would do anything that we needed help with. The only direct contact of a continuing nature we had with the Naval Academy was to transfer some funds from ONI for Severn River Naval Command, since we had to be paid by them, and they, of course, had no budget to cover our activity. So ONI transferred funds annually from the ONI budget to the Severn River Naval Command supply officer or comptroller, and he would issue the paychecks for our civilians.

"At Y1, were you working directly for the DNI? In terms of the content of your work, there was no intervening command? Did anyone at NSA check your work?"

"Absolutely not. We worked directly for the DNI, as we had been part and parcel of ONI. In fact, we still showed the Y-1 organization with the Y-1 designators on the ONI directory, the ONI Roster, with NFOIO simply put in parenthesis after my name. So, for all practical purposes, we were still ONI. But the important thing was, and this became significant later to other people, the important thing was that the personnel in Y-1 no longer counted against the headquarters establishment of OPNAV. It became significant because it led people to discover that you could create field activities and move them out of the immediate area, or out of the Pentagon, and thus reduce your OPNAV manpower ceiling while still having the organization in place, effective, and working for you. Since we were the first field activity of ONI, and we were created because of our geographic separation, it became a forerunner of what is now effect, field activities of ONI."

"And you also provided a template for how other offices in the Pentagon to do such things?"

"Right," he said with a slight grimace. "We are responsible for the sprawl. Little by little other activities were moved out of the Pentagon to the Hoffman Building in Alexandria and out to Suitland and were created as field activities of ONI, and subsequently NAVINTCOM was created to be the umbrella over all of these field activities representing the DNI."

"Was being at Fort Meade any more of a handicap in getting your needs across and getting your support to the DNI?"

"No, it was no handicap because I continued to live in North Arlington. I started out each day from Arlington and returned to Arlington each evening, and, if need be, I could start my day in the Pentagon or finish my day in the Pentagon and not be grossly dislocated because I could simply change my schedule. In the beginning, most of our other officers also lived in the Virginia area, where we had all been in order to work at Arlington Hall.

Some of the people relocated their households out to the Maryland countryside or to

the Laurel vicinity. There were no quarters available on the Post. Many people continued to live in Northern Virginia. I found it quite convenient to do so because I would, whenever necessary, spend a half a day in the Pentagon, either in the morning or in the afternoon or conceivably all day. But the wildest commute was that done by Captain Bill Hatch. He was determined to raise his family away from the big city and owned a farm out in Leesburg.

He would drive to Arlington long before dawn and we would car pool to Ft. Meade." I gave a low whistle. "That might be the worst commute in Washington," I said.

"And remember: it was before the Beltway, which didn't get finished until 1964, and by that time I was on my way to Pearl Harbor again. Remind me to tell you about what the commuting was like one of these afternoons."

"Yessir." I asked Peter for the check and counted myself lucky that I might have made my last drive to Fort Meade. At least on official Business, anyway.

SPACE AGE

• • •

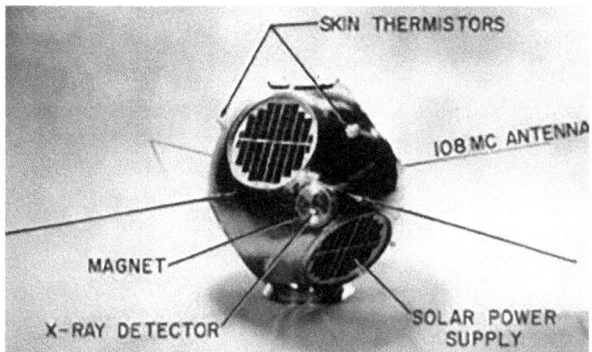

GRAB RECONNAISSANCE SATELLITE, CIRCA 1961. ONE OF THEM HUNG FOR
YEARS IN A STAIRWELL AT THE OFFICE OF NAVAL INTELLIGENCE.

It was winter, and that meant that the sky outside Willow was
dark as we arrived, and the little vestibule between the inner and
outer doors to the bar side of the restaurant was welcome to prevent
the chill wind from blowing in.

It seemed to be a Wednesday, for some reason best known to
itself, and at the office across the street there were five proposals in
progress with the capture guys screaming at the Proposal Managers,
the day workers being expected to contribute deathless technical
prose in the hours after the day-job was wrapped up.

There is nothing I hate quite as much as proposal writing, and
it was hard to change gears and just talk to Mac. I actually had a

decent notebook with me, for a change, since the big IT conference in Detroit that summer had provided fancy wire-backed hardcover books that we could fill up with all sorts of interesting information. I brought it with me and was proud to be halfway prepared for a session with Mac Showers.

"So, last time we talked you were the head of Y Branch in the Pentagon. Can you tell me what was going on in 1961?"

"Space Age. We had a significant role in the development, launch and operations of the Navy's first ELINT satellite activities."

"There is a big exhibit I saw out at the National Reconnaissance Office about GRAB and POPPY vehicles. That was amazing that the Navy was first into orbit for satellite recce. POPPY was still in orbit when I joined the Navy almost forty years ago."

"There was no NRO then as we know it today. GRAB—the Galactic Radiation and Background collection satellite—had been launched by the Naval Research Laboratory early in 1961 and was actually orbiting and collecting ELINT over denied areas. I think this may still be classified because this satellite was launched under the guise of a solar radiation detector. It was openly advertised satellite for the purpose of collecting solar radiation measurements."

"Did it?"

"Yes, and very well. But, more importantly, it also collected ELINT from the Soviet landmass."

I took a long refreshing pull from my glass of Happy Hour White. It was particularly crisp—Willow uses the low special hour price as a loss-leader to get people to drink more and order from the bar menu. "Electronic Intelligence, like radar and other signals the Soviet ships would put out while operating on the high seas." I remembered the old watchword from the Ocean Surveillance System: "Live by ELINT, die by ELINT. "In my day, we had a thing called "hull-to-emitter-correlation" where we could actually identify the exact ships by their unique frequency signature."

Mac nodded and took a sip of Anchor Steam beer. He was in a better mood than I was since the Docs had given special dispensation for him to have a real beer for a change. "Over time, several of these satellites were launched and did collect intelligence effectively, and the results were analyzed. They turned out to be a very productive platform, and I'm certain they were the first ELINT satellites owned by anybody of any country."

"Admiral, I was a budget guy like you at the end of my career, and wound up funding all sorts of strange things. How were the satellites paid for if they were a clandestine program? Was "Y" Branch a money manager as well as a substantive material and program manager?"

The Admiral looked into space for a moment, collecting his

thoughts while appearing to examine the rows of Willow's fine wines on the shelves behind the bar. "We probably had a hand in justifying the need for the collection in order to obtain the money, but it was Navy Research Lab money and we probably assisted, as I recall, on that as well as other projects. We helped NRL justify the need for the funds, but the money came out of research and development capital. Of course, there was the issue of divided chains of command.

NSGA SUGAR GROVE'S BIG DISH. IT IS FOR SALE.

Since NRL was a laboratory, it reported up to the Assistant Secretary for R&D. That whose pot of money paid for the that project, plus the development of what we called the Big Dish at the Naval Security Group station at Sugar Grove, West Virginia."

"I have been there when it was active and Byrd was pushing for more government activities to move out there. We had to respond to questions about the alternate Atomic Clock, and the FBI fingerprint labs I heard it was for sale to private owners once Senator Byrd was no longer the Chairman of the Appropriations Committee."

"Bobby Byrd was a magician in getting the bacon home to West Virginia. We tried to support other schemes, too, and spent many hours briefing people at the Pentagon and the government. The money was coming from Navy R&D to NRL for the construction of the Big Dish, which was to be a moon-bounce ELINT interceptor, a moon-bouncing communications ELINT and communications interceptor using the moon as our satellite, so to speak."

"It is only fair, I suppose. Sugar Grove used to advertise that it was in a radio silence' zone due to the mountains and the terrain. No interference from other terrestrial signals."

"That's how it was advertised by NRL. But it never panned out because the cost escalated to the point that it was shelved. It started out being about a $150 million dollar project, and then it crept up to $160, $170, $180 and began getting shaky when it went over $180, and as it approached $200 million, they killed it."

"I hate it when that happens," I said. "And that was in 1961 dollars. That was a pricey capability."

"You mentioned Senator Byrd. There were some political aspects of that too, with all that money flowing into West Virginia; it tended to have some inertia to folks on the Hill."

"I hate to think back on the number of things that went to West By God over the years he was in the Senate. I think a good deal of money has since flowed into West Virginia for similar purposes, but that didn't happen on my watch. The "Y" Branch role in the NRL solar radiation satellite was to perform the tasking."

"I remember the fights to get targets in the Soviet Far East into the collection deck for the imagery satellites. How did it work for Y Branch, translating technical collection requirement into instructions on the bird in orbit?"

SENATOR ROBERT BYRD, PERHAPS THE MOST EFFECTIVE PRODUCTIVE PORK PRODUCER IN U.S. SENATE HISTORY.

"We had an outfit called "TOG," which stood for "Technical Operations Group." It was a very simple acronym, and was supported by the Naval Security Group, NRL, the Navy's master engineer Howard Lorenzen, by the Science and Technical Intelligence Committee (STIC) in the form of Bill Howe, and by "Y" Branch's Fred Welden before me, and then eventually myself. We met once a month to develop the tasking schedule for the on-off times for the sensor package to optimize collection on what we were interested in. We were restricted in what we could do, since to protect the secret of GRAB's real mission, we couldn't operate the satellite full-time when the bird was over the Soviet landmass. We had to operate it on a random and part-time basis so that we didn't disclose the fact that it was collecting against certain targets at certain times."

"Yeah. I was riding in a bus in an exercise and they parked us over a highway overpass to hide when the Soviet Satellites were scheduled to come overhead."

"Yes—vulnerability to satellite reconnaissance, or SATVUL times. So the purpose of TOG was to work out the schedule and submit it to the operators of the satellite and the intercept stations who were copying the downlink at the Naval Security Group stations at Bremerhaven and in Japan."

"This was science fiction stuff for the World War Two generation. Were you guys supposed to be experts on orbital mechanics and the operating characteristics of the satellite?"

"We did indeed. I even went down to Cape Canaveral and watched one of the launches. That was the first time I'd ever seen a launch. But I just recite that as one of the interesting things that made my year as head of "Y" Branch absolutely fascinating. One of the finest short tours of duty I ever had. But, early in that tour, and I'm talking now in late 1961, within a few months after I reported, I was called in by Rear Admiral Vernon L. Lowrance, who had become DNI, and he asked me if I thought I could handle the job of Fleet Intelligence Officer, CINCPACFLT."

"What an opportunity to serve in Eddie Layton's old office! What did you say?"

"My answer to Admiral Lowrance was that I was way too junior, I still only had a year in grade. I thought I was too junior to handle an assignment like that."

"Funny about that. Your Boss in WWII, Admiral Nimitz, said the same thing when they picked him to go receive Husband Kimmel after the attack on Pearl."

"You are quite right on that. But a Chester Nimitz only comes along every couple generations."

I nodded in agreement. Mac finished the last of his second beer, and announced that he needed to get to the dinner service back at The Madison across the street.

I waved at dark-haired Serena behind the bar to see if we could get the check. Mac got off lightly since the night Liz-S awarded the admiral an official document saying he would henceforth be drinking for free at Willow. It is a much better deal than I am ever going to get! Old Jim has calculated our monthly tabs, and he claimed we pay about $1,400 a month in rent for the stools on which we sit, with alcohol included, of course. Thank goodness, it is *not* in 1961 dollars.

BLUE RIBBON PANELS

• • •

Willow was just starting to take on the energy of the post-work Happy Hour Crowd. Liz-with-an-S was behind the bar, and working some of the questions for admission to the Virginia Bar. She is already admitted to the New York and New Jersey ones—but that wasn't quite good enough to crash the protective gates of the Lawyers Lobby here. We were pulling for her to get out of Food and Beverage and get back to being an officer of the Court the way she had intended—and maybe whittle down some of the student debt that she had amassed getting through Law School.

I had given her a set of flash cards I found on the web to drill on likely questions she would encounter in the exam-and take a new direction in her professional life. In between tulip glasses of Happy Hour White, of course.

Old Jim was seated in his usual place at the apex of the Amen Corner. He was a constant fixture at Willow, at least when he was not involved in a boycott resulting from overfamiliarity, since he lived with his bride Chanteuse Mary just down Utah Street and with his bum leg, it was just about the right distance for him to travel. Even Uncle Julio's up the block was a little far to hike. Anything else—The First Down or The Front Page, for example—were a cab ride away, much less the bright lights of Clarendon, the next stop on the Orange Line route into town.

Mac was prompt and nicely turned out in a jacket and sports shirt. I had no idea what I wanted to cover in this chat. When we had talked about his time at Fort Meade with the Naval Field Operational Intelligence Office we had got to the point of talking about one of the first major controversies at the national level of the Intelligence Community-the amount of money we spend to do things like miss the collapse of the Soviet Union. That sort of thing.

But this was before the trauma of the Vietnam conflict, which I hoped to get to in the next few weeks. I picked up my pen, now that Liz-S had me set up with a full glass of wine and a ready supply of cocktail napkins. "OK," I said. "You were just getting to the first of the big Blue Ribbon Panels in which you participated. I remember you mentioning the Schlesinger Panel, the Church and Pike Commissions and all the rest. What was this one called again?"

"It is a recurring phenomenon in the Government. Something

400 ••• *Vic Socotra*

happens, everyone agrees it looks bad, then they call in some smart people and make recommendation that may or may not be helpful. Early in my tour at NFOIO, our Director Admiral Frost became involved with what became known as the Robertson Committee," said Mac. "Mr. Robertson was an Assistant Secretary of Defense— and was charged by the Secretary of Defense to find out what communications intelligence was costing the United States government."

"That has got to be a huge amount," I said. "The whole NSA and all the activities the Services had embedded in their structures. I doubt if anyone could track it all down."

"Robertson tried. This was the beginning of serious budget concerns in the intelligence and the COMINT business. So the Robertson Committee was formed of Army, Navy, Air Force—the intelligence agencies and the cryptologic agencies of Army, Navy, Air Force—and NSA, and it was monitored by State Department and CIA and other people in the government who used COMINT. It became a large committee that worked for several months to price out costs of COMINT, to find out where there might be duplication, where there might be wasted effort, where there might be gaps. Their ultimate outcome was to come up with the first consolidated cryptologic program and in fact, still exists."

"But this was before Vietnam. Before the major expansion of everything and they already were wondering what they got for their money?"

"1958, I think things really got rolling, or late in 1957. Shortly after reporting as chief of Yl in March 1957, Admiral Frost called me into his office in the Pentagon to ask me to Bak—stop him as a member of the Robertson Committee. As Assistant Secretary of Defense, he had been charged by SECDEF to convene a group of authorities and, over time to examine and price out the total cost to the Defense Department of all the cryptologic activities then ongoing. Admiral Frost was a member of the group as the DNI, and I was his backup. He probably had backup from the Naval Security Group as well since that is where the equities were located. It was a large committee: Army, Navy, Air Force, various elements of the Defense Department, the budget people, of course, and the Armed Forces Security Agency, or National Security Agency, which it probably was called by then. We even had State with at least observer status, the CIA, and the other elements of the government who were users of· the cryptologic product. It was a large committee, and it consisted of the committee of Principals and then there was a working group for those of us who were the backup would meet to do our spade work."

"Sounds like the Deputy's Committee where real work gets done. You never get anything effective with just the grownups there."

"Well, we sat for several months, certainly the better part of a year. I don't recall the full extent of it. but the result was the creation of the Consolidated Cryptologic Program (the CCP), which since then is still the cryptologic program that carries the budget program for all cryptologic activities in the Defense Department. In those days, it was managed then by the Director, Defense Research and Engineering within the Pentagon. DDR&E was the Pentagon point of contact for all cryptologic activities under the responsibility of the Secretary of Defense, and DDR&E was, in effect, the office of the SECDEF which was designated as being responsible for the activities of the National Security Agency. I mention this because it was the birth of the CCP and the first time that anyone had attempted to put the cryptologic budgets together into one package, which happened then and has continued since and has also been applied to other types of activities and programs in government."

I do recall that the total bill was the magnitude of the program, and it turned out to be an astounding figure. This surprised everybody. I don't recall what it was at that time, but, when it was all put together and added up, it was an amazing number. That effort took a good deal of my time because this committee or the working group met in lengthy sessions at least weekly, if not more often, in the Pentagon. I was at Arlington Hall for the first nine months of that tour, so that was not too much of a dislocation, but after we moved to Fort Meade in December 1957, this increased the commuting time and the commuting activities.

I, however, lived in Arlington at that time so, if there were meetings to be attended in the Pentagon that were going to go all day or a good portion of the day, I could go from home to the Pentagon or I could go from Fort Meade to the Pentagon and then come home from there. So it wasn't too inconvenient for me. That's all I'm going to say on the Robertson Committee, since there were things that I still can't talk about."

Was there anything else you wanted to know about my time at NFOIO? I think we've talked about the rest of my tenure as Y1, or more formally OP-922Y1. I wrote about it for the Naval Intelligence Professionals, but that was probably before your time as editor."

"I will track it down, Sir. Should be interesting reading with a little perspective."

"The article I wrote pretty much covered the tour of duty there, including the organization and functions of the office, some of the things we did, and our relationships with the National Security Agency. I didn't go into those relationships in detail because they might include some subjects that would still be classified. In fact, I *know* they would. In a classified session, there may be some things that would be worth reporting as a result of that—particularly our role in working with the National Security Agency."

"Let me see what I can find regarding the article. If I have questions, I am sure we can sort it out over an Anchor Steam Ale and some Happy Hour White."

"I would be delighted. I just have to get the Doctors on the same sheet of music as Willow."

Liz-S came down the bar with a flash card. "Does Atkins v. Virginia strike any chords?" she asked. "The defense apparently relied on a single exculpatory witness for a capital murder case."

Mac pursed his lips. "I think I actually remember that one," he said. "No Blue Ribbon Panels, though."

"No," said Liz-S. "The question is about the execution of mentally retarded persons and whether it is "cruel and unusual punishment" prohibited by the Eighth Amendment."

"It may be cruel," said Mac. "But in my experience, it is hardly unusual."

HIGHER EDUCATION

• • •

PANORAMIC VIEW OF THE NAVAL WAR COLLEGE CAMPUS
AT NEWPORT, RHODE ISLAND.

Willow was quiet that afternoon. It was a bit early for Old Jim to limp down the block for happy hour, and Mac Showers and I had were the only customers inside. A couple of tables were full out on the patio, but the sultry August weather had everyone at a low ebb. If there were ceiling fans inside, the dust motes would have danced in the breeze. it would have been a setting out of any classic film noir movie. Mac was in an aloha shirt, no jacket, and I was still in my work clothes. We were talking about the early 1960s, times that we overlapped in this world and which, on a good day, I vaguely recall. Behind the bar was Tex, the former Marine who stands about six two and has put on a little weight since he got out, but is always jovial and usually seems happy to see us. Tex was an interesting guy. He has some sort of a sinus problem that causes him to snort once in a while, and a cheerful indifference to the niceties of some social conventions. Which led to an unusual transaction at the bar one afternoon that will have to remain between the two of us.

But I was not there for commercial activity, regardless of how gratifying. I wanted to get through Mac's time in the Y1 organization and get back to the Pacific, where I had some major questions to ask. So, without much ado, I launched right into it with my trusty pen in one hand and a glass of almost-Happy Hour white wine in the other. "So, with part of your organization co-located with NSA, you must have had a lot of interaction with The Fort."

Mac nodded in agreement. "There were many interesting liaison activities that I conducted as Y1 with NSA, some on very friendly and beneficial terms, not always at loggerheads with them, like the submarine program seemed to be. One at which I am not prepared to go into detail was the attempt between the Navy Research Laboratory and ONI on the one hand and NSA on the other to try to understand and capture the Soviet short-signal burst transmission—which was just coming into use in the late '50s, and surprising to say, we did succeed with great satisfaction in doing that. That was when I first met Dr. Gene Fubini, who NSA called in by NSA as a consultant to try to understand the electronic mechanics of the signal. I appreciated from the outset that Gene was a genius in these matters, and he and I subsequently became very close friends in many endeavors that the Navy had underway. He, of course, took up positions in the Pentagon—DDR&E, etc.—and I might comment that he was also is my neighbor, just around the corner when I was still in the house in North Arlington."

"I remember the di-graph for the classification, Admiral. So, you were still doing SIGINT in Y1?"

"We did in the organization, yes, absolutely. You could not get away from it."

DR. EUGENE FUBINI, THE LEGENDARY NAVY ELECTRONIC WIZARD.

"That makes sense since Naval intelligence has always been focused to a large degree on being able to locate shipborne radars and communications. It was our bread and butter and still is.""Okay, let's move on to my detachment from Y1 in the summer of 1960, relieved by Captain John Q. Edwards, who has subsequently written in the Quarterly about some of his experiences in Y1. I was selected to attend the senior course at the Naval War College in Newport, which I did.

"I love military higher education," I said with a laugh. "Were you a Commander or Captain?"

"I had been deep-selected for Captain, but I only made my number days before being detached from the Pentagon. I was happy to arrive at the War College in Newport as Captain rather than as a

Commander, which was useful from the standpoint of getting better quarters and accommodations. I looked on my year at Newport as a sabbatical after all hard work and long Pentagon hours. It turned out not to be so relaxed, because I became so enamored with the academic community, and I got reintroduced to a delightful library where I could go and contemplate the things I had seen. The library at the War College is open 24 hours a day. You can go there any time to read or research, or whatever. Each student, of course, was assigned the writing of a term paper. My term paper was on the general subject of "Soviet Submarine-Launched Cruise Missiles," which were just coming into the inventory and in testing stages at that time. So I did a research paper on those for which I fortunately got an outstanding grade. I had some help from my old shipmate Art Newell, who was the intelligence specialist assigned to the War College faculty at the time, and coincidently was also my faculty· advisor. But the War College turned out to be truly a delightful experience."

"I never had to worry about early selection. But I agree. I felt the same way about the Industrial College when I finally got a chance to go to school. Who would have imagined getting full pay and allowances while getting your master's degree?"

"The War College was not a degree-granting institution then. They worked out something with the Salve Regina College across the road to accept the War College curriculum as meeting their standards for a degree later, but it didn't exist then. Higher military education was certainly a change of pace for me and the family. We were at Newport when the Bay of Pigs raid occurred."

"1961, right? That would have been the Cuban Crisis period," trying to get my recent history straight. "The Cuban Missile Crisis was in 1962. The Bay of Pigs was right after JFK came into office, which would have been January of that year."

"Yes. I was at the War College at the time of the Bay of Pigs. John Edwards was in Yl, because he mentions some of this incident in his account. I also was at the War College when the two Brits defected."

"Burgess and McLean, right?"

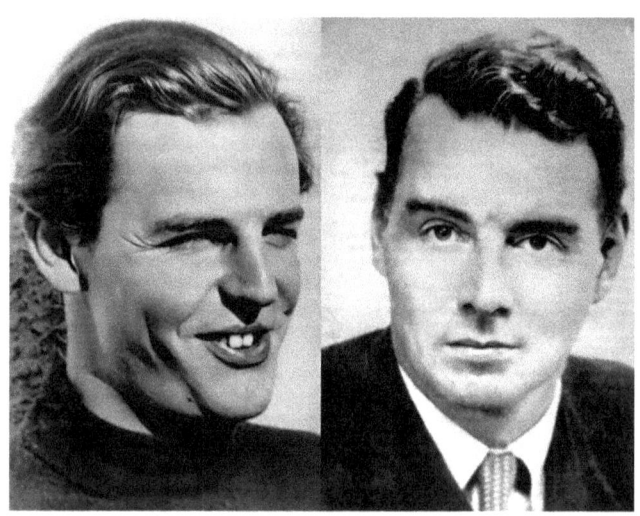

BRITISH TRAITORS GUY BURGESS AND ALISTAIR MACLEAN.

Mac gave me a thin smile. "At the time, I was absolutely astounded to read The New York Times about the Soviet accounts of the debriefings in Moscow after they arrived in Moscow. That was really the first time that U.S. and U.S./British SIGINT activities were openly made known to the Soviets, as far as we knew."

"We seem only to keep secrets from the people who pay for them. The Soviets had moles everywhere, and there was not much they did not have. Down through the years, we have certainly had a lot of people willing to give things up to them, either out of ideology or for cash." Mac nodded gravely.

"Yes, it was all printed for the general public. They were absolutely astounding accounts. I was sitting up in Newport reading about all this, knowing I was going back to the Pentagon to head "Y" Branch and deal with what had been exposed to the Soviets. Anyway, the year at Newport was interesting. That was the year that John Kennedy was elected. One of our speakers at the War College was Paul Nitze, who at that time was campaigning for John Kennedy. I remember his talk about naval activities and naval operations. During the question and answer period, there had just been an incident where a Soviet SIGINT trawler had attempted to recover a missile being tested off Cape Canaveral. And the question was asked of Paul Nitze, who was preaching the more aggressive nature of military operations that would come into being during a Kennedy Administration.

THE HONORABLE PAUL NITZE AS SECNAV.

One of the questions was, "If you were Secretary of the Navy, what would you have done about that SIGINT trawler that attempted to recover the missile?" And he shot back, "I would have sunk the son-of-a-bitch!" And he got a standing ovation. Two or three hundred students in the auditorium, very impressive then, and of course, Nitze later *did* become Secretary of the Navy. I don't recall him sinking any trawlers, though."

"The OpNav Staff must have told him there was a State Department?."

"Right. Among other things. Anyway, on completion of my War College work, I was ordered back to the Pentagon to relieve Captain Fred Welden as head of 1Y Branch. I reported in August 1961 and was delighted to be there. I felt that I had finally arrived at the epitome of OPINTEL in the Navy and was in the billet I had long admired and sought after.

"This was in the D-ring on the fifth floor of the Pentagon?

Mac nodded. "The D-ring on the fifth floor. The area was blocked off at both the 7th and 8th corridors, I believe. We had an enclosed hallway in which we had the parts of "Y" Branch that were not at The Fort: Y-2, Y-3, and Y-4 .

"What did each of the groups do?" I asked, taking a sip of wine and letting the pen rest on the bar where it was safe.

"Well. Y-1, or the Naval Field Operational Intelligence organization was still at Fort Meade, under John Edwards. Y-2 was responsible for briefings to the Chief of Naval Operations . . .."

"What we knew as CNO Intelligence Plot. Jake tried to detail me there one time."

Mac nodded, "We called it that, too. I'm trying to distinguish between Y-3 and Y-4." His face was screwed up in thought. "Y-4 was ELINT and was responsible for the peripheral collection activities, mainly of the VQ squadrons—the electronic collection P-3 Orion aircraft. This was really the Washington contact point over the activities of the VQ squadrons, and Y-4 was mainly manned by aviators that had been in the VQ squadrons and understood the avionics of flying those airplanes. Y-3 then, I can't recall. It might have been collection management."

"You had this whole empire?"

"Yes I did. The whole shooting match, and I should be able to recall. It seems to me that Y-3 had something to do with collection activities involving the U-2 because, on this tour of duty for the first time, I was briefed among other things on the U-2 operation and also, before the tour was over, on the advent of satellite operations. They were collecting photography of military installations in denied areas."

"Which would have been useful before Gary Francis Powers was shot down over the Soviet Union. That is about as denied as you can get."

Mac smiled. "That was why we moved things onto low earth orbit. Not that many Russians there in those days."

"And now we get our rocket engines from them."

"It is a strange world, isn't it?" asked Mac as he contemplated his Virgin Mary.

WAY DOWN YONDER (IN VIETNAM)

• • •

BOOMER BEHIND THE BAR AT THE AMEN CORNER.
PHOTO: SOCOTRA

Willow was bustling, and it was with good reason. The last Friday night of the month was the fabulous Buffalo Night, in which Tracy O'Grady re-creates the most famous sandwich to ever come out of Buffalo, New York: the astonishing 'Beef on Weck.'

BOW is composed of a slow-cooked locally raised grass-fed steamer round of beef, thin sliced and piled high on one of Kate Jansen's amazing Kemmelweck Kaiser rolls, topped with fennel and sea salt. They are so good I normally ordered two in advance to prevent being disappointed when they sold out. I would normally eat one at the bar and take the other one home to snack on the rest of the weekend.

Mac normally dines at The Madison across Fairfax Drive where he lives, but he could sense the excitement as the Regulars filed in and asked Big Jim behind the bar whether the "beef was resting" prior to being carved up into succulent slices.

Mac was dressed in a natty coat and tie, since Buffalo Night we all come straight from the office and look vaguely professional. Boomer was behind the bar, and bantering with the regulars as she normally does in the brassy manner that endeared her to us.

I picked up my pen and slid a stack of napkins over to where I could get at them. "I am afraid we are getting to your return to O'ahu in the early 1960s."

"Why do you say it that way?" he asked, having a sip of Virgin Mary since the Doctors were being cruel to him this week.

"Because I am starting to remember some of the events you are describing and it doesn't seem that long ago," I said with a laugh.

"You are not going to pin me down like that. I will tell it the way it actually happened, and to do so I have to go back a year before I took the family to Hawaii. It was 1961 when I returned from the senior course at the Naval War College in Newport. After graduating in the summer, I was ordered to ONI again to be the Head of "Y" Branch."

"Which was the Special Intelligence Division, right?" I asked. "It had become something else by the time I was around."

"Yep. We had a suite in the Pentagon for the various components, not far from Rear Admiral Vernon L. Lowrance's office as the

DNI. Shortly after I arrived and got settled in the building, I had a session with the Admiral in which we discussed my future career. I told him that I would like to branch out of Naval Intelligence to do some things other than pure OPINTEL to broaden myself. I told him I was in a rut and I would like to try something different like an attaché assignment or some other form of Naval Intelligence service."

RADM VERNON L. LOWRANCE AS DNI. A WWII SUBMARINER, LOWRANCE
WAS AWARDED THE NAVY CROSS FOR GALLANTRY. PHOTO: U.S. NAVY

"Sounds like fun being a diplomat and open-source intelligence collector going to cocktail parties, but a lot of Attachés got passed over later for being away form the Navy too long."

"I know, but I was restless. The DNI was sympathetic to my request, but he said "You're in a rut that an awful lot of people would like to be in with you." And then he said, "I don't have a lot of sympathies for your desires to go elsewhere."

"I told him, if I was to stay in that line of work, that my ulti-mate desire was to get Eddie Layton's job as the Intelligence Officer for the Pacific Fleet. That was my desire, and had been ever since the war. My time in Europe and the growing realization of the Soviet Threat never kept me from keeping a fond eye on the activities at CINCPACFLT and who got the jobs as the top Naval Intelligence officer in the Pacific, which was viewed as a job that could be a springboard to selection for Flag. I wanted a shot at it, eventually."

"You can imagine my surprise when just a few months later, the DNI called me to his office on the fifth Deck and told me he was prepared to send me to CINCPACFLT in the near term if I really wanted to go. I had some misgivings. I only had two years in grade as a Captain, and the job was really for a mid-to-senior officer. I questioned whether I would really be acceptable to the Fleet Commander being that junior, and he assured me "that's no problem, I'll take care of that.""

"Always take a challenge" said Mac, as Boomer came by, inspecting the level of liquid in everyone's glass. "That was my motto, and I told him, "Yes, I'd love to go. So, I agreed, he agreed, and orders were cut to detach from the Pentagon in the summer of 1962 and relocate to Pearl Harbor, where I reported in August."

"Just in time for the World Series. Didn't the Yankees beat the San Francisco Giants?"

Mac smiled. "I never had much time to follow baseball, and there is a six-hour time difference, remember. It made listening to the games too hard, and who cares about the Yankees, anyway. I had a big job to start. I really was the assistant chief of staff for intelligence when I reported. Three years later I was selected for admiral, and I made my number in December of 1965. That promotion made me the senior Division Chief on the staff for my last six or seven months in the job."

"How did you feel about that? Your staff during World War II was fighting a global war, was all over the Pacific, and Eddie Layton was about the only Captain that you dealt with. Did you feel that there had been a grade creep out there at that point?"

Mac furrowed his brow. "Well, I think there was a grade creep everywhere. But, I think it was a necessary evil since Admiral Nimitz had kept his PACFLT staff during the war a very lean organization. The number of officers in the Intelligence Division could practically be counted on the fingers of one hand. But there was the huge organization of JICPOA and the Fleet Radio Unit to back up the Intelligence Division. After those large organizations were done away with in 1941, the intelligence staff itself became larger. And that was true, more or less, of all the divisions on the staff. The staff grew in numbers, because it was an creature unto itself, rather than being supported by a number of subordinate organizations. I don't remember the numbers in the Intelligence Division, but we must have had 30 or 35 officers in the various sections. It was a large and very busy organization." Mac paused to wet his whistle and munch on a stalk of celery that stood proudly in his glass.

"Of course, the Vietnam War was just cooking off, and there were many things that had to be done. I mentioned outside organizations that supported us. We did have the Fleet Intelligence Center over on Ford Island to support us, but FICPAC was of relatively little direct support to the staff. FICPAC instead was supporting the fleet, mainly with photography work and with producing targeting materials.

Those kinds of things could be produced on a production line basis. What we were doing at the Headquarters, where I was responsible for figuring out what it all meant with analytical work and current reporting and responding to the daily crisis, whatever it was."

"That hadn't changed, Sir. Same deal, different year." I looked at my notes. "The first American casualty of the war was in the 1950s, I said taking a sip of Happy Hour white. "But you were there when things were just starting to get hot. When I worked there, we still had your Vietnam work-week: half day on Wednesdays and they expected a half-day in Saturdays, too."

WILLOW'S BEEF-ON-WECK COMES ON THAT FABULOUS KAISER-STYLE KEMMELWECK ROLL, PIERCED BY A GARNISH OF THREE DEEP-FRIED OLIVES, A SIDE OF CARAMELIZED ONIONS AND TRACY O'GRADY'S TRADITIONAL CREAMY HORSERADISH SAUCE. THE ONE I TAKE HOME COULD MAKE THREE OR FOUR WEEKEND LUNCHES.

"We felt we needed the continuity on the mission and didn't want to lose details if the analysts were off the whole weekend. We were busy. We were setting up field activities in the Philippines. In hindsight, I think we responded to Fleet requirements one at a time. It was like this In 1961, when I was back in "Y" Branch in

Washington. We had commenced a series of peripheral intelligence collection patrols along the Soviet and Chinese coasts known as the DeSoto Patrols." Ed Nielsen and I originated DeSoto patrols, but I am tired and that is going to have to wait to another session."

Boomer came down the bar and slid one of the fabulous beef sandwiches in front of me. Mac looked on with interest as she topped off my wine.

"There are other important issues to talk about right now. Do you think the Beef on Weck sandwiches are sold out yet?"

CRUSADES

• • •

Mac had some back problems last week, but he came roaring back and was ready to hit the bar at Willow on Monday. The weather was loosening up just like his sacroiliac—unseasonably warm in Arlington, and the rising temperature featured the best of both worlds—the ladies shed their coats to revealing advantage but kept their tall leather boots with the spike heels.

"I was admiring the view on the way over," I said. Mac had beat me to the bar by minutes, and he was sitting by Old Jim and Mary, who stopped on her way back from the office downtown.

"There was a cracked rail on the Metro," she said. "Things were a mess all day."

I nodded, grateful that I do not have to travel far to get to the office. I looked around for a pen, found that I had conveniently forgotten mine, and borrowed one from Katya, whose dark-eyed beauty graced the business end of the bar along with Tinkerbelle and Jasper and the lovely Liz-with-an-S. "So," I said to Mac, grabbing a stack of napkins, "Where were we?"

The door to the bar swung open with a rush and in walked Point Loma with Jiffy, another Midway sailor in tow. I knew this was going to get complicated, particularly with both Johns, with and without H's, and The Lovely Bea.

Mac is a babe magnet, for sure, and he was in fine fettle and thoroughly enjoying the first of his two Race 5 India Pale Ales. He cleared his throat and said, "I don't know. What do you want to talk about?"

"Well, the big news is about Hawaii, and where we should stay if we go for the 70th anniversary of the Battle of Midway this summer."

"I don't know," he said. "The Hale Koa is all the way downtown, and the events are supposed to happen at Pearl Harbor."

"I like the Rainbow Tower of the Hilton, too. That is where the Navy used to put us when they tented the houses to kill the monster cockroaches. The housing area at McGrew Point was built on landfill, and there was no way to really eradicate the things. They just moved from house to house."

"I was always in quarters at Makalapa, on the rim of the Crater. It was nice to walk to work."

"I bet. Kimo is out there now, in your old job as the Fleet Intelligence Officer. It will be good to see him again back in his element. And Admiral Paul, who works up at the PACOM HQ at Camp Smith. This will be fun if we can pull it off."

"We will see," said Mac. "Paul was leaving on official travel to Thailand, I think."

"Should be time for COBRA GOLD," I said, thinking of the best joint naval exercise in the world, since it normally came with a four-day port visit in Pattaya Beach. Colonel Ike was further down the bar, huddled with Jake. I pointed at him, saying: "Ike just got back from Cambodia. I have always wanted to go there." Katya topped up my white wine. "And Laos, of course. Damn, there is a lot to see."

Mac smiled. "I was one of the last Americans on the Plain of Jars," he said. "That was the trip with Lt Gen Bennett when I was Chief of Staff at DIA. We were visiting the Ambassador, G. McMurtrie Godley, which made things confusing since we were both known as "Mac." Everyone else knew him as "The Field Marshall," since he was involved in everything going on in the country, political or military."

"So, you were there just ahead of the Pathet Lao guerrillas?"

Mac nodded. "It seemed like a good idea to see the place while we could."

"And now we can again," I said. "I sort of feel like just heading west from Hawaii this summer."

Mac took a sip of beer. "I think I have been to SE Asia for the last time," he said. "That was the same trip we saw Admiral Rex in Saigon, and had dinner with him and Admiral Bud Zumwalt."

I picked up my pen. "Wait," I said, scribbling. "That is impossible. He was not Zumwalt's Intelligence Officer. I have talked to the guy that relieved him the year before. Rex was back here, working collections issues."

VADM Earl "Rex" Rectanus in Vietnam.

"That may well be, but when General Bennett and I walked into Zumwalt's quarters, he was there, big as life. I sat between him and Bud."

I screwed up my brow in puzzlement. "If he was there, and I believe you, Sir, that means something had caused him to be sent temporary duty from Washington, and it must have been something big that he did not mention to you."

"Like the case of the missing Jack Graf," said Mac. "But we have been down this rabbit hole before."

"Jack's loss was a major crusade for Rex in his later years. I learned a lot helping to research the available evidence on his

POW-Missing in Action status. A lot of the Naval Intelligence guys obviously followed the case pretty closely, and they came close to rescuing him at least once, with the camp where they held him showing signs they had only left hours before."

"Torture can make anyone talk," I said. "I heard the SEALs even found some of the Viet Cong interrogator's notes."

CDR GEORGE "JACK" GRAF ON HIS SECOND IN-COUNTRY TOUR
AS A NAVAL INTELLIGENCE LIAISON OFFICER-NILO.

Mac nodded thoughtfully. "That is why the whole shot-while-trying-to-escape and Jack's body being buried in a place where the river washed it away is an interesting story."

"Yeah. When I found out that Jack had been to the Kodak School to learn about how they were going to do electro-optical imagery from earth orbit before he went back to Vietnam I was stunned. They never let people with those clearances get far out of Saigon for fear they would be captured and compromise the biggest secret in the Intelligence Community. And then Jack parachutes down right into the middle of them after he got shot down."

"Do you think Rex was in Saigon to do a damage assessment on his loss?" asked Mac.

"I don't know, and if Jack was traded to the Soviets, we have lost our window of opportunity to find out from the KGB files."

"I don't suppose we will ever know the answer, but to get a technician who knew how the spy satellites really worked would have been worth a lot to the Russians."

Mac shrugged. "Case closed, as far as the POW-MIA folks are concerned. But it would explain why Rex was there. The Navy would have been embarrassed at the screw-up that put Jack in a place where he could be captured."

Then we drifted away from mystery, and talked about other ones, cancer being one of them, according to my notes, and then about Mac's top-ten recipes. "Eggplant Parmesan, hands down," he said. "I did all the cooking for the last few years that Billie was still living at home. I got pretty good at it. The stuff they serve at The Madison is abysmal. They don't have a clue."

I scribbled frantically. "I need the recipe," I said. "I would like to try it. What else did you have in the rotation?"

"Chili con carne," he said. "I have a recipe I invented myself. Spaghetti, apple crisp as a dessert."

"No pear pies, like the ones from the C-rations on Guam."

"No, definitely not. I don't think I have had a pear of any kind since the War. And tenderloins. I would get the big ones form the Commissary—I would toss one in the over at 400 degrees for an hour, then turn it off an let it rest for an hour. Couldn't miss. Perfect every time."

"That sounds delicious," I said.

"The kids liked a thing we called 'Porcupine Balls.'" "That doesn't sound very appetizing," I said.

"Actually quite tasty. We used a pressure cooker. Dangerous things, and you had to watch them closely. I would take hamburger and shape them into meatballs mixed with regular white rice. When they cooked under pressure—I don't recall how long, but not too long—the rice stuck out like the quills on a porcupine."

I wrote it down. There were several other conversations in progress. Point Loma was talking about Ops Officers he had known on *Midway*, and Mary was saying why Bob Ryan the weatherman had changed stations, and why he got eased out of his old job at Channel 7.

The threads were all interesting, and I decided to stop writing and concentrate on the wine. Mac smiled. "Good, now that you are not writing things down, I have a story for you that you can't tell."

I put down my pen. "I am all ears," I said.

It was an interesting story, and it is too bad I can't tell you. But I promised. Life is interesting, you know? And like the Jack Graf story, it doesn't always make a lot of sense. "What did Shakespeare say about life?" I asked Mac.

He smiled broadly. "A tale full of sound and fury," he said.

I gestured at my notes. "And told by an idiot," I said. "Who would be me."

DESOTOS

• • •

BOOMER TAKES THAT CRITICAL FIRST HAPPY HOUR ORDER. PHOTO: SOCOTRA

The people and employees at Willow seem to blend together into an ersatz, if quite real, family. When things were going well at the restaurant, all was right with the world. When there was a shake-up in the personnel line-up behind the bar, it was like losing part of the family.

Peter, Big Jim, Tinkerbelle, Liz-S, Boomer, Tall Sammy, Briana, Marvin . . . the departure of each had special circumstances and a certain period of mourning.

The bartenders were the highest visibility members of the crowd, of course, and Old Jim who holds court at the Amen Corner of the long bar. Boomer was still working there and had not moved down to Shooter McGee's on Duke Street, and she was big and bold as life there at Willow this afternoon.

Mac and I had made plans to meet and talk about technical intelligence collection operations in the Vietnam conflict. He was still off alcohol (on Doctor's orders, dammit!) and just sipping a bit of his first Virgin Mary about halfway down the bar when I slid onto a stool next to him.

We exchanged pleasantries but didn't talk about the election, the economy, or the prospects for World Peace breaking out. I had my pen, and I actually remembered to bring a notebook. No napkins this afternoon, and I looked at the Admiral expectantly, and he knew just where to start.

"It was 1961. In order to cope with expanding threats all over Asia, we had commenced a series of peripheral intelligence collection patrols along the Soviet and Chinese coasts known as the "DeSoto Patrols.""

"Ed Nielsen and I originated the concept back in Washington before I went out to Hawaii. It was a valuable concept that got us unique intelligence on all sorts of emitters, communications and military training. In fact, using the Desoto platforms soon had them on station from the Bering Straits to the Java Sea and through the Malaka Strait. We were trying to patrol and collection in limited areas from the Arctic Circle to south of the Equator."

"For example, when the Indonesians acquired SAM-2 GUIDELINE missiles from their cozy relations with the Soviets, we mounted DeSoto Patrols off the Indonesian coast to try to intercept the electronic signals from the missiles. That was some of the first SAM-2 missile ELINT that the United States ever obtained. Unique stuff and very useful."

"We used to have pre-configured vans we could load on ships. Did you use the existing electronic warfare fit on the destroyers, or did you have them take on board special equipment?"

"No real difference except for the sophistication of the equipment. We put a hut on the limited deck-space available on the destroyers to house special. Our concept for the DeSoto Patrols was to take ships from the fleet, add equipment that the Naval Security Group would configure, and NSG would put a team onboard to work inside the hut. The intelligence collection mission was accomplished entirely from the hut by the NSG Ship-riders. They were just augmentees to the regular crew for the length of the patrol. It worked well."

"It sounds just like some of the operations we conducted off Nicaragua with the USS *Sphinx (ARL-24)* when I was at Third Fleet, I said. "They pulled her out of the inactive ship facility at Bremerton to refurbish and patrol off the coast during the Contra war."

"Sphinx was with us in Vietnam, too. Want to know how the program got its name?" "You were interested in defunct American cars?"

"Close," laughed Mac. "When I was head of Special Intelligence Branch at ONI in the latter part of 1961, we received a message one morning from COMSEVENTHFLT."

"I am proud to have been a 7ᵗʰ Fleet sailor, Admiral. My son was, too, after he did his time out there in Yokosuka."

"I quite agree. But you asked about where the name came from. In those days, Navy messages were had a standard slug to put in the subject line. In the case of this message stream, the subject line was: DeHaven Special Operation Off Tsingtao.

The 7ᵗʰ Fleet Staff was proposing a special collection effort by USS *DeHaven* [DD-727], with NSG shipriders to operate off the People's Liberation Army base at Tsingtao because there was some unusual training activity happening there. Ed Neilsen and I read that message early one morning in the Pentagon, and I said to Ed, "This is going to be voluminous; this operation which will create a ton of message traffic. And among, other things, we can no longer deal with that long slug line every time we have to write a message on the subject."

"So, I took my pencil and I underlined the initial letters "DeHaven Special Operation off Tsingtao" and I abbreviated it, "DESOTO." That, I guarantee you, is the source of the name for Desoto Program, which ultimately became extensive, successful, and eventually controversial."

"*Totally* controversial after the attack on the *Liberty*, not to mention the loss of the *Pueblo*."

USS *LIBERTY* (AGTR-5) RECEIVES ASSISTANCE AFTER SHE WAS ATTACKED AND SERIOUSLY DAMAGED BY ISRAELI FORCES OFF THE SINAI PENINSULA ON 08 JUNE 1967. AN SH-3 HELICOPTER IS NEAR HER BOW.

"That is what takes us back to the Vietnam War, finally. It was a Desoto Patrol that created the Gulf of Tonkin incidents. *Turner Joy* (DD-951) was on a Desoto patrol in the Gulf of Tonkin when she was allegedly attacked by North Vietnamese torpedo boats.

"I have wandered into the continuing academic debate about the Gulf of Tonkin incident and I am pretty sure there are still parts of me missing. Now, you say "allegedly." Most people feel there is no doubt about the first of the two incidents, but obviously there is considerable controversy over the second."

Mac gave a grim smile, "I say "allegedly" because of the

controversy. I am personally convinced there were two attacks, but, I say "allegedly" so I don't get in trouble with the True Believers who maintain there was only one. Life is too short for that."

USS *TURNER JOY* (DD-951)

"Amen," I said, putting down my pen and raising my now-empty glass to get Boomer's attention.

"Bottom line," Mac said, "was that SECDEF McNamara reviewed the purpose and efforts of the Desoto patrols in view of the *Turner Joy* incident and issued an edict that, henceforth and thereafter, DeSoto Patrols were terminated. The program ended because of the Gulf of Tonkin incident—and I thought at the time it was a rather short-sided view. The DeSoto program had been extremely effective. Sure, *Turner Joy* ran into a little trouble, but that was essentially one of the purposes of the program. What do you smart computer guys call it? 'Not a bug, but a *feature?*'"

"I am a Luddite, myself, Sir. But I take your point. Presence has a certain imperative all its own."

"Exactly. We were operating with a warship near the periphery in order to stimulate reactions and activities for intelligence collection purposes, and it worked like a charm. But here is where things went wrong. We stopped using real warships for a variety of reasons and went with platforms that had more deck space, were cheaper to operate, and were not needed on the Gun Line in the South China Sea."

"That was part of the Soviet intelligence collection scheme. They labeled some big auxiliaries as "AGI's," or Auxiliary General Intelligence. They were unarmed but awesomely capable of sucking every electron out of the spectrum, and they could hang around ports like Pearl for weeks and weeks."

"We bought into the concept that our collectors should be unthreatening. We decided to use dedicated auxiliaries as collectors instead of warships. We configured the *Pueblo* (AGER-2), *Banner* (AGER-1), and *Palm Beach* (AGER-3) as essentially unarmed platforms that would be non-threatening and non-provocative to continue the Desoto-type of intelligence collection. You remember what happened as a result of that. The North Koreans were able to pick off the *Pueblo*, steal the radios and crypto gear, seize the ship and hold the crew hostage for eleven months."

"It was a mess. I understand that some people say the point of the capture was to give the radios on *Pueblo* to the Russians, which allowed them to reverse-engineer the devices so they could use the keying material that John Walker started providing them around the same time. Having both the device and the key code gave them access to our tactical communications in Vietnam. It was still going on as late as the 1980s."

"If we had to fight them at sea, we would have had a worse situation than the Japanese when we penetrated their

JN-25 communications," said Mac. "If you have no secure communications, the enemy knows your intent."

I allowed myself a little shiver. "I was out there then, and that would not have been pretty."

Mac shook his head in agreement. "I was part of the *Pueblo* damage assessment later at DIA. I wish that had been one of the key conclusions—that our communications had been massively compromise and we might have changed the radios sooner. I will have to tell you about it, and how I met Wanda, who wound up in the Front Office at DIA for the next thirty-odd years."

"I have always respected her," I said. " She was always kind and always professional." "And remember, it was the age of the mini-skirt," laughed Mac.

"I just wish I could have met her when you did, Sir!" Mac just smiled.

APRIL FOOL'S DAY 2010

• • •

ADM Thomas Moorer in 1964. Photo: U.S. Navy

I t was April Fool's Day, 2010. I saw in the morning traffic that the town of Topeka, Kansas, had decided to rename itself "Google" for some reason best known to the Mayor and the city council. Mac Showers was feeling a little housebound and was available to chat, and on a splendid day in the northern Virginia Spring, he decided to fire up the champagne-colored Jaguar saloon and drive across the street to the only saloon I have, the fabulous Willow Bar.

My only prank in honor of the spirit of the day was to show up at the office and pretend to work. Things had been busy in that curious way of the contracting business. One moment we had five proposals going to try to win work at the Defense Intelligence Agency, and then they were all submitted and most of us were still walking around with slack jaws and fatigue in our eyes. It is a strange business, and totally dependent on a bunch of GS-13 government drones.

I had a copy of a picture that Mac had given me when we were talking about the DESOTO patrols he had developed and the events that surrounded the Gulf of Tonkin incident.

"Admiral You showed me a photograph a few minutes ago of your Situation Room at CINCPAC. Admiral Tom Moorer was there, you were there, and a whole bunch of other people were hanging out. It is obviously off-duty hours, because one officer is in a very Hawaiian shirt. You said that it is a picture of the night that you were all waiting for word from Washington after the Tonkin Gulf incident. I wonder if you could give me some more context for all that—the things I have heard all concentrated on the events in the Gulf, not the reaction of the U.S. Government and the Navy staffs that had to take action in response."

Mac was wearing his aloha shirt under his sport jacket, an affectation that he preferred to demonstrate that he had spent the better part of a decade working in Pearl Harbor. He took a sip of Virgin Mary and said: "You'll have to verify the date that the picture was taken, but I do recall that it was the date that we mobilized the staff in the evening that the reporting about the incident was available

"I looked it up, I must have been around August 8th, 1964, right?"

"That certainly would fit. That was about mid-tour for me at PACFLT, and Admiral Moorer was the commander-in-chief."

"Mr. Rumsfeld told us we couldn't say that anymore. His position was that there was only *one* Commander In Chief, and he lives on Pennsylvania Avenue. I still can't keep it straight."

"Mr. Rumsfeld was a character even the first time he was SECDEF in the Ford Administration. We knew that President Johnson was going to attempt to get the authority to order retaliatory strikes against North Vietnam, so we pre-planned that the staff would be mobilized and we'd all be present in the Op Center that evening. We had all the staff division heads there. The gentleman in the aloha shirt was our assistant chief-of-staff for logistics. The Plans Officer was there and the Operations Officer was there with all of his officers. I was there with one or two assistants, and all other key members of the staff were present. The picture was taken at the time in which we were maintaining our Seventh Fleet and air status board awaiting the word from Washington to launch the strikes. That word did come during the evening in Hawaii so it obviously was late at night in Washington. The strikes were launched, and the President, as I recall, wanted an immediate report on the effectiveness of the bombing. So, we stayed there, essentially all night, until the aircraft were recovered and we were able to provide reports back to Washington on the effectiveness of the strikes."

"I was always a Bomb Damage Assessment guy. They must have used Viggies to get the post-strike imagery. Were the targets selected at COMPAC . . . hell, CINCPACFLT, then?"

"Yes, I think we selected the targets, coordinated with Washington, of course, but I think we nominated the targets and carried out the strikes. This type of mobilization in the evening for these strikes obviously didn't continue. I think we did do it two or three times in the beginning because we didn't know what kind of reaction we would get. We knew the North Vietnamese were not able to react against our fleet units, but we didn't know who else might react against our fleet units. So, we were prepared for any kind of reaction.

That's why we had the full staff mobilized and ready to do whatever was necessary. Over time, obviously, as these strikes continued and were repeated, they became routine and it wasn't necessary for the PACFLT Op Center to be fully mobilized. Instead, it was done by the Seventh Fleet and carried out routinely."

"So, the night of the picture you still that feeling of, "Crap, here may go the start of the war." That must have been fairly alarming?"

"Vic, you are a master of understatement."

"That's your third war?"

"I missed the Korean War in that I was in London. I was about as far away from that as one can get. I was just as happy about that. It was an ugly one."

"You don't count the Cold War as one of them? I do."

"Point taken. But really, the one that counted was 1941-45 for me."

"Trust me, Admiral. You didn't miss a *thing* in Korea."

WHAT NEDZI KNEW

• • •

REP. LUCIEN N. NEDZI, 1969 FROM CONGRESSIONAL PICTORIAL DIRECTORY.

Mac was just back from the beach on the annual Shower's family reunion, and I was just returned from the First Congressional District of Michigan, and it was clearly time to get together at Willow and talk about what happened to the Intelligence Community in the Ford Administration. He was intimately involved in all of it from his new perch on the IC Staff at CIA.

It was the organization that acted in the CIA Director's community role as the pater familia of the fractious agencies allegedly under his charge. In that role, he was known as the "Director of Central Intelligence," not the Director of CIA.

I know, I know, small difference semantically, but large in terms of authority.

It was 2011. In the decade-long series of interviews, we had arrived at that strangest part of American political life, or at least the strangest until Donald Trump's candidacy.

I was interested in the Michigan congressional delegation, since I could use my parent's address in The Little Village By the Bay for voting purposes, provided I didn't also try to vote in Northern Virginia. I had not gotten around to changing things since I had voted there (absentee) for most of my military career.

"Trooper Bart" Stupak was my Congressman from Michigan's First District then,, a ruggedly handsome former State Patrol officer. He was a Democrat, and a relatively conservative one as reflects the philosophy of the First District, who are country working people, though he went with his party on 96% of his votes.

At the end of his elected time, he came in conflict with the leadership of his party on the healthcare debacle, largely over the issue of Choice, which was a hot button Up North, and ultimately impacted his decision to leave the Congress. I saw him dozens of times when I worked on the Hill, and met him on at least one occasion. It is not that unusual to run into celebrities here in town, or at least the political versions.

FORMER PRESIDENT JERRY FORD AS A WWII NAVAL OFFICER.

When I marched up and introduced myself as a constituent, he seemed vaguely alarmed.

His unease probably stemmed from the fact that folks from the second-largest Congressional district east of the Mississippi get down to the nation's capital get down to bother him. The First, as you know, encompasses the entire Upper Peninsula, and depending on the population level, a good chunk of the lower one, too.

That includes the Trolls who live in the Little Village By the Bay. I say "trolls" because that is what the independent-minded residents of the proud Upper Peninsula (the "UP") call all of us who live "below the bridge" at Mackinaw. For our part, the only things in the UP worth knowing are Lake Superior, the pasties, iron ore, moose, wolves and the odd bear.

And snow, of course. That goes without saying.

We had our favorite sons. Jerry Ford was a Michigan grad and a Navy vet from Mac's war. He represented Michigan's Fifth District for a quarter century, the one with all the block-headed Dutch in it. His last eight years in Congress were as the Republican Minority Leader in the House when that seemed like a permanent position.

He was a reliable enough apparatchik to be pulled onto the Warren Commission whitewash of the Kennedy killing, remember?

When Betty Ford passed away it brought the Ford Administration to a final end, with many of those loose ends still unraveled. My son and I went down to the National Mall the night after Jerry passed, and his funeral cortege made the pilgrimage in front of our vantage point near the Washington Monument, passing his former homes in Alexandria and the national sites on the mall.

Between the Warren Commission, the pardon of President Nixon and the conquest of the Republic of Vietnam nine months into his unelected Presidency, you would think he would be a more controversial figure than history has cast him. But I say the hell with it.

He was a good congressman for us in Grand Rapids, when the Socotras lived there, and people genuinely liked him.

During Ford's time in the Oval Office, foreign policy was characterized in procedural terms by the increased role Congress began to play in that sandbox and by the corresponding curb on the powers of the President. That is what ultimately brings me around to the matter of Lucien Norbert Nedzi (born the year after my dad).

He was the Democratic Congressman from the 14th, later the 1st, between the special election in 1961 and the time he threw in the towel in 1980 and did not seek re-election.

Nedzi may be the last guy in a position to know what was going on then, besides Mac. Nedzi was born in Hamtramck, the Polish enclave surrounded by Detroit. He was a Wolverine, like President Ford and me, and went to the U-of-D law school in between stints in the Army in the Pacific War and his recall for Korea. Bummer for those guys who got called back to active duty from their civilian jobs and new families. My Dad would have been recalled, but all the Reserve squadrons east of the Mississippi were devoted to the Russian threat in Europe and stayed home.

CIA DIRECTOR WILLIAM COLBY

I think that is why Mom laid down the law and made him get out.

Getting recalled to Asia to face hoards of Commies was a tough blow for a lot of GI's and sailors, but they just sucked it up and did what they had to do.

Like I said, Nedzi was elected as a Democrat in 1961 after

Thaddeus Machrowicz died in office. The key to his being remembered at all is the curious thing that happened in May of 1973. Lucien had the unique position to be a member of the House Armed Services Committee, and Chairman of the Special Subcommittee on Intelligence.

Director Colby at CIA had inherited the toxic "Family Jewels" from Jim Schlesinger which detailed the activities conducted by the CIA that might have "exceeded the charter of the Agency," in the delicate terminology of the day.

Colby was summoned to discuss the matter with Nedzi on the Hill in an extraordinary session detailed by a declassified Memo-for-the-Record drafted by CIA Inspector General William Broe. I asked Mac about it, and he beamed in remembrance over his Virgin Mary at the long mahogany Willow bar. He even had a copy of it, printed off years later when it entered the public record.

The two-hour session included a discussion of the full report, item by item, including the sensitive portion that remains redacted to this day.

According to Broe, the congressman found it found sobering. Some of the topics included:

a. *Alien documentation furnished to the Secret Service. Nedzi desired more information concerning the reason why issued, the use, and how controlled.*

b. *Financial support to the White House in connection with the replies to letters and telegrams as a result of the President's speech on Cambodia in 1970. He requested more information on this subject.*

c. *Beacons furnished to Ambassadors. He was interested in the number issued to Ambassadors and the position the State Department took on the use of these beacons. He was interested if the Department of State was pushing this program, as he believed they should be.*

d. *Logistics' acquisition of police equipment. He questioned whether LEAA, Department of Justice, should not be doing this rather than the Agency.*

e. *He noted Logistics furnished telephone analyzers, and desired to know what they were and how used.*

f. *[redacted]*

g. *OER's crash project concerning Robert L. Vesco requested by the DCI. The Congressman was interested in who outside the Agency instigated the project and why was it stopped.*

h. *Several ORD projects indicated research done without knowledge of the host system or on unwitting subjects. He was of the opinion that this was risky and recommended it be terminated. He stated he would like to see a directive go out to the researchers concerning these practices.*

i. *John Dean's request re Investors Overseas Service. He reviewed the six reports that had been furnished. He noted, however, that the item stated "there were multiple channels to the Agency from the White House" and requested information concerning these channels.*

j. *Alien passports. Mr. Colby advised that he planned to review this whole subject and the Congressman agreed with the need to do so. The Congressman asked Mr. Colby if the Agency had considered how much of the information just reviewed with him could be made public. Mr. Colby stated this had not been done yet, and spoke to*

the question of sources, methods, and the impact on the institution.
The Congressman stated that in the current climate he felt it was
necessary to open up more information to help clear the air.

Mr. Colby stated the Agency would give the matter deep consideration, and added he had been thinking of a general statement
along these lines to be used at his confirmation hearing.

The cat was about to come out of the bag—or at least part of
the cat out of one of the bags. Tip O'Neil acquiesced when Nedzi
demanded to be placed in charge of a Select Committee to formally
"clear the air," a position he assumed in February, 1975, and which
he abruptly resigned in June.

I wondered about that, since Otis Pike replaced him as chair for
a tumultuous and shocking set of disclosures about what had been
going on for years. I will be interested to see what Mac has to say
about that. I wonder if Nedzi found out something about the IC that
he did not want to have on his permanent record. After all, he was
in a position to suddenly be investigating the work he should have
been doing as the senior oversight official on the IC.

I know this: in the Congressman's remaining three terms in
Congress, he never got closer to the intelligence community than
the Joint Committee on the Library. I am not going to speculate
any further, though.

There is a way to find out, though. Like Trooper Bart, Nedzi
didn't return to Michigan when his congressional time was done.
He went into the lobbying business and lived right here in next-door
McLean, and his number is in the phone book.

If he is not in the same boat as my dad, he might have been able to answer a few questions if I called him up. You never can tell. Thank God we had Mac to ask. He leaned over one time and whispered, *sotto vocce*: "LBJ did it."

I just nodded and made a note on the cocktail napkin.

THE HATCH WAY

• • •

LT BILL HATCH, MAC'S CARPOOL BUDDY ON THE DAILY
PRE-BELTWAY SLOG TO NFOIO IN PRE-BELTWAY DAYS.

I t was a curious coincidence. I was going to a funeral on my
birthday. CAPT William Hatch had passed away on the on the
6th of June, the sixty-second anniversary of the landings at Normandy.
When he died, he was the oldest living naval intelligence officer.
Now it was our pal Mac Showers, and as our friendship deepened,
I offered to drive him to the ceremony out in distant Leesburg.

I did not know him personally, but I edit a journal that is devoted
to the study of the profession, and I felt an obligation to provide a

presence from the organization. People of his vintage are getting scarce, and it is imperative to honor them before their families and keep the memories alive while we can.

The memorial service was held at Saint James Episcopal Church out in Leesburg. Bill Hatch and his lovely wife Nancy had purchased a working dairy farm out in Loudoun County back in 1950 when the memories of the Civil War were still alive, and Leesburg was remembered as the place of the great victory against the invading Yankees at Balls Bluff, just outside town.

There is a postage-stamp national cemetery there, home to the dead of that day long ago. It is now in the middle of a suburban subdivision, and the last time I was there, the Confederate battle flag was flying from a staff in the middle of it.

The battlefield would have been pretty much the same as it was when the Yankees were hurled back into the river when the Hatch's and their five children moved in. It was a different Virginia then.

Mac and I navigated out the Dulles Access Road, and then the Greenway Toll road. We made excellent time from Arlington, and I wondered how Bill Hatch had done it. He did not retire until 1973, and there were years of the commute from the farm down to the Pentagon on two-lane blacktop roads.

Mac laughed and began to tell me some commuting stories that made my toes curl. "We used to drive across Constitution Ave to get to Ft. Mead. Bill would come in from Leesburg, we would jump in one car in Arlington and race for The Fort."

Living in Loudoun County was a statement in those days, when sensible people thought that the site of Dulles International was only a little bit closer to town than Mars. Admiral "Shap" Shapiro worked directly for Bill in Washington when he was the Deputy Director of Naval Intelligence, and he said once that try as he might, he was never able to get to work ahead of him, and that was after he had milked the cows, done the chores and driven down to the Pentagon.

It was a pleasant day for a drive. High thin clouds, blue above, a little cool for a day in early June. I took Business Route 15 into the historic district of the city. Leesburg is the County Seat, and the courthouse and downtown buildings square off across narrow streets. They were filling up with weekend tourists as I passed through. I had excellent directions from the Church website, and pulled up in front with plenty of time to spare.

St. James' Episcopal Church is a massive structure of stone hewn from the bedrock of the county. It has expanded over the years, with the present main church laid down in 1895. Shelburne Parish traces its history back much further, to colonial times, when it was carved out of the western portion of the Parish of Cameron in 1769, when the Church of England was still the established state religion. The first church in Loudoun County was built in 1733, and called the "Chapel of Ease for the comfort of the people above the Goose Creek."

Before 1769, the Chapel of Ease was served by visiting ministers. The first resident rector was the Rev. Dr. David Griffith, a distinguished churchman who served from 1771 to the year of the Declaration of Independence in 1776. His service lasted past the surrender of Cornwallis at Yorktown, and the disestablishment of the Anglican Church. Three years after independence, in 1786, he was the first man elected bishop in America.

Unfortunately, funds for his travel to England could not be procured, and he died, unconfirmed.

Captain Hatch had a lot of friends. Mac Showers is a former Deputy Director of the Defense Intelligence Agency who served with Chester Nimitz in the war in the Pacific. He made an eloquent statement the day after the news began to circulate of Bill's passing. That was basically what I knew about him, except for the excerpt from an ancient Blue Book that he was the senior living intelligence officer.

Now it is Mac.

Mac said that he had been in declining health for some time, but remained at the farm south of Leesburg. His eldest daughter had informed him that Bill me died peacefully with his children in attendance.

I walked up the street to the formal entrance to the church. Bob Juengling was standing on the porch, looking up and down the street to see who was approaching for the service. Like me, he was there to honor the memory and service, and to sit in the back of the church. Bob is a little older than me, and he retired with 33 years of service. He would have stayed longer, too, if they had let him.

We chatted on the porch, making ourselves useful by opening the door for the family and friends arriving, including Admiral John Marocchi and his wife. They seemed a bit frail, but still independent. John is now the second oldest surviving officer, and one of the last with World War Two service.

We like to be useful. We greeted people and gave directions

and information from the porch, and held the door for folks with canes. A Navy Captain in dress blues strode briskly up to the steps. I did not recognize him, and when I introduced myself, he said he was a local reservist, and had seen the notice in the paper. He said it was only appropriate that someone was there, in uniform. It was the least he could do.

Bob told me what he remembered about Bill, between opening the door for people and saying, "Good morning!" He had been an icon in our community. A gentleman and a family man who, with his lovely wife Nancy, set a standard for Naval Intelligence couples to follow. There was a generation of "Hatch-trained man," and Admiral Shapiro considered himself one of them. His first assignment under him was in the Soviet (Y) Branch in the office of Naval Intelligence, which Bill headed, and in 1968 he was his deputy at the U.S. Navy headquarters in London on North Audley Street.

I will always remember that building fondly, and the pubs around it. It was where Ike Eisenhower had his headquarters before he crossed the Channel in 1944.

In London, Bill had been the intelligence officer for Admiral John Sidney McCain, the father of Senator John McCain. The Admiral became Commander in Chief of the Pacific Fleet after London, and it was then that his son was shot down and captured by the North Vietnamese. Some say the only reason that John got enough medical treatment to live was the fact that his father was the Admiral in command of the naval forces off the coast.

We passed the minutes with small-talk as the time for the service grew closer. The breeze was fresh, and smelled of rolling fields and rich grass. With about five minutes to go, we entered the church and greeted the family in the vestibule, introducing ourselves and saying that we were there from the old organization to honor Bill.

Inside, the church felt massive and secure. Flags from the State and the Parish hung from the rafters, and a few others. The one across from our pew at the rear of the church was of plain red, with five white stars. I realized with a start that this was the Church were General of the Army and Secretary of State George C. Marshall worshipped.

Bill Hatch was in good company indeed.

The program contained a rich biography which I scanned as the celebrants assembled at the rear of the church.

William Nagel Hatch was born on June 20, 1918, in Stockton, CA, or as it was formerly known, Muddville, in the San Joaquin Valley of California. His Mother and father were both second generation Californians, and the family had a connection with the sea: they owned the Monticello Steamship Company, which ferried passengers between Berkeley, Vallejo, and San Francisco.

His father died in England while serving on active duty in the Navy in the great influenza pandemic at the end of World War I. Bill never met his father. His mother later married another naval officer, a nautical engineer.

With the marriage came two stepbrothers, and the life of a Navy family, moving between Mare Island, Pearl Harbor, and Portsmouth in Virginia. Bill met his future wife Nancy there, though his stepfather's career had him graduating from Ionlani High School in Honolulu.

Bill was off to college when the Japanese bombed Pearl Harbor, and he immediately dropped out to join the Navy. He began his career as an underwater demolition expert, helping to clear the mines laid by German Unterzeeboots at the entrances to the Chesapeake Bay and Charleston Harbor.

A young man with obvious potential, he was selected as a '90 day wonder,' attending a three-month commissioning program at the U.S. Naval Academy.

He and Nancy married at Tampa's Saint Andrew's Episcopal Church on July 22, 1943. Bill was then deployed on a U.S. minesweeper with a crew from the Russian Navy, home-ported at the northernmost Soviet Naval base in Murmansk. Duty in those frigid waters meant certain death if the ship went down, and the Germans prowled the arctic sea to sink the merchant ships in the convoys, trying to shut the vital lifeline that kept the Soviets going through the dark days of the Nazi invasion.

Based on his EOD experience and language skills, Bill was transferred to the U.S. Embassy in Moscow. When the war ended, Bill was one of the original group of Naval Reserve officers to be selected for transfer from the Reserves to the Regular

Navy. In 1946, he was designated a Special Duty Officer-Naval Intelligence.

Bill's knowledge of Russian and his wartime experience in the Soviet Union qualified him as the Navy's first designated Soviet Navy expert, and the Soviets were quickly changing from wartime ally to implacable opponent. His initial assignment in this role was with the Navy Security Group at the Nebraska Avenue Security Station, where the Department of Homeland Security now resides.

In 1950, Nancy and Bill decided to purchase the Mill Road Farm, a working dairy concern in rural Loudoun County. Nancy came from a Navy family, too, and based on their vagabond upbringing, they wanted some stability for their children. It was an island of stability, as Bill's career took the five children to London, England, Naples, Yokosuka and Newport.

He had begun his connection to this church shortly after purchasing Mill Road Farm. He was on the Calling Committee for the Rector, Rev. Frank Moss, to whom he confided: "I have five heathens, do something with them."

The inevitable occurred. Having identified a requirement, Bill rose to meet it. By the end of 1954, Bill was superintendent of St. James' Sunday school, Nancy was a member of the Altar Guild, and the children had all been baptized.

The farm was always in the middle, and that is where Nancy and Bill retired in 1973, the same year I graduated from college, his professional life concluding and mine just beginning.

I finished reading as the organ began to swell and the formal service began. Rev. John Ohmer, Rector, strode down the central aisle with dignity and an immense strength. His hair was cropped short, his body powerful, harking back to a robust sort of Christianity that used to be practiced in these parts. His homily spoke of the time, five years ago, when Nancy passed away. He said he had a hard time keeping a stoic face when he arrived at the farm. The connection between Nancy and Bill was so deep, so profound, that the sorrow of being present at the parting made him want to weep.

He said Bill took off his wedding band, and asked him to make sure that it went with Nancy to the funeral home. "Till death do us part," he said. And that is when Rev, Ohmer lost it and began to cry himself.

As for me, I waited until they sang the Navy Hymn at the end. It always gets me. I can't help it.

Mac I and I hung out in the Fellowship Hall, and he paid his respects to an old shipmate. When we drove back to Arlington on the Green Way to the Dulles Access Road and eventually to I-66, I marveled that this is what a commute was like, in the times before the Beltway and all the inter-state highway system.

Those guys were *tough*.

GREEN EYE-SHADES

The sun was golden on the patio at Willow, filtered through the abundant greenery of the trees overhead. It was so pleasant a July day that all the tables were occupied and Jasper the bartender was running around in his manic manner, keeping the patrons well lubricated.

Mac was wearing a jacket and tie due to a medical appointment but looked cool but proper. I was wearing a jacket and clip-on bow tie, having strolled over from the office, and was happy we were in the cool darkness of the Willow's long bar.

"We were talking about the transition from being at DIA in uniform and joining the Intelligence Community Staff as a career civilian. What was that like? I noticed the cultural change immediately when I went to Langley for my last tour in the Navy."

Mac cleared his throat and reached for his Anchor Steam Beer. He had bullied his doctors into permitting him to have a couple beers, an act of courage for which he was quite proud.

"When I got to Bolling in 1966, I got tagged with the Program and Budget nonsense. DIA had been charged to do build and present to the Congressional Committees a consolidated strategic intelligence budget between Army, Navy, Air Force and their subordinate elements."

"That was Major Force Program 3, I said. I had to learn about all that crap when they sent me to be the budget staff director at Bolling. The Services still had to do their tactical intelligence— Program 2—submissions through their parent services. Everything else strategic came up through our staff."

Mac nodded. "Yes. Title 10 and all that. Well, the Congress expected the Director of Central Intelligence to do the same thing for the whole community, and that is one of the reasons Bronson Tweedy had his eye out for me. He thought I knew what I was doing."

"Apparently they were under the same illusion about me," I said shaking my head. "It was mind-numbing, and I don't think I really ever did understand it." The happy hour white was a *pinot grigio*, crisp and refreshing. I had a new pen, a Pilot G-2 that made incredibly fine letters on the napkin in front of me.

"It may seem odd nowadays, since the establishment of the office of the Director of National Intelligence, but Congress really only thought the IC Staff was about the budget, not policy. Once I reported to Langley as a civilian, I got involved in the program and budget review work. My account was NSA. They initially considered having me be responsible for the DIA portfolio, but they thought better of it."

"We called it the General Defense Intelligence Staff when I was at DIA. We found the Community Management Staff—your old IC Staff—to be irritating busybodies." I glanced down at the bar where Liz-with-an-S was attending to some self-important-looking gentlemen in suits and caught her eye to request a refill.

Mac took a sip of beer. "That is the nature of Washington, to be irritating. It was decided that since I had just come from being chief of staff for DIA, that this might be a mistake, since I had just been there doing the budget. Plus, they thought I might not be able to work effectively if I was put in a position of conflict with the Agency where I had been the Number Three. Because of my long-time association with SIGINT and SIGINT activities, they gave me the NSA account."

"That is where my former Deputy Senior Executive Jerry wound up. He was a holy terror in oversight for the Fort. They feared him. Big job. Lot of money in play."

"I was extremely comfortable because I had many friends at NSA, and from my time working at Ft Meade in the Naval

Intelligence Field Office, knew how their system worked. I was an advocate of many things that NSA was doing, but I also was a critic of what wasn't going that well. So, I felt good about handling the NSA account."

"Who got DIA," I asked. "I had once assembled a list of all the GDIP Staff directors, and posted pictures of the ones I could find in the office in the Pentagon. All got lost in the 9/11 attack. The jet almost hit the staff offices head on. We were lucky the whole staff wasn't killed like the people in the spaces in the next Wedge."

"A good day to not be in the Pentagon. A young Air Force vet named Marty Hurwitz got the DIA account," he said.

"Marty! I actually know him. He must have moved down from Langley to take the job. He was there for, like, 20 years. A real pro, and a good friend to the Commands outside DIA when the Agency didn't want to play nice."

"If you want a friend in Washington," said Mac with a droll smile.

"I know. I ought to go to the pound. I am glad I am not doing that anymore."

"I didn't really care for that type of work. And even though I enjoyed my new role with the IC staff and my work with NSA, I still never really liked the program and budget cycle and the ridiculously complex process arduous nature of what the U.S. government forces itself through each year.

"It certainly taught me more things I didn't want to know than any other job I have had." I rolled my eyes thinking about the tiny interior office in the Pentagon and the endless hours at the desk or in meetings.

Liz-S clearly recognized my distress and topped my glass off. "There, there," she said smoothly. "No bad eyeshade people are going to come and take you away."

"Thankfully, I only had the NSA account for two cycles. Even though I did it reasonably well, I never felt comfortable."

"It was way out of my comfort level, and when you talk to people about the endless racking-and-stacking drills about requirements and available cash to meet them, their eyes cloud over."

"With good reason. What happened next?"

"Well, in the latter part of 1973—after Bill Colby became DCI and Sam Wilson became the Director of the IC staff—there was a new problem the community had to deal with, and they needed a go-fer. I kind of became a jack-of-all-trades and a floating representative on the IC Staff, as did Jack Thomas."

"General Thomas was there when I arrived at Langley twenty years later, I said with a start. The General was an amazing guy, quiet and kind, and he must have had more than a half-century in the business."

"At least. He was one of the first Air Force general officers who was not a rated pilot. I managed to get out of being a go-fer after a couple years, trying to damp down the mistrust of the IC Staff. People thought we were scheming to undermine the authority of the Agencies. But then something else came up. In 1975, I became a special assistant to Sam Wilson who, in effect, was working with Colby to set up a special project. It is still around and still working. But that is one of the things we agreed not to talk about," said Mac. "I still have friends up there and they might not approve hearing about it from the Willow."

MAJOR GENERAL JACK THOMAS, USAF-RET.

Old Jim put down his Budweiser long-neck firmly on the bar at his place at the Amen Corner and growled that he swore he would never disclose a thing, so long as we quit talking about the budget.

I think Mac and I were both happy to do that.

WHIM-ICKS (AND OFFICE POLITICS)

• • •

A WWMCCS CONSOLE IN THE EARLY 1970S. THE OLD JOINT STAFF
COMMAND CENTER IN THE PENTAGON BRISTLED WITH THIS SPACE
AGE EQUIPMENT WHEN I ARRIVED IN 1990.

It was oral history night at the Willow bar. Owners Tracy O'Grady and Kate Jansen operate a place with style, and some well-dressed people were thronging the long bar after work.

Well, there is them and us, and Old Jim is not above barking at them if they get too close to him. I had my weapons of mass dissemination laid out on the bar: pen, tulip glass of the loss-leader happy Hour White, and a fresh pen. "We are not going to be able to dive too deep into your "post-retirement" career at the Central Intelligence Agency," I said, restating the groundwork. But can you give me an overview of what was happening then?"

Mac was bright and alert and the oncologists had backed off when he confronted them about the desire to have a beer once in a while, and was enjoying an Anchor Steam carefully poured by Big Jim, the Pittsburg bartender and Willow's current liquor manager.

Mac looked at the rich amber in the glass, the perfect foam topping, and the dancing motion of the bubbles throwing themselves toward the surface like reverse depth charges.

"Shaaah," said Jim, and he lumbered up the bar to see if Old Jim needed another Bud.

"I would prefer not to go into some parts. But I think there are some broad-brush topics and personalities we can talk about all you want.

"Works for me. I don't want to get into some beef with the Agency at this late date. But you were removed from active duty in the Zumwalt purge of every flag officer that had been senior to him when he became Chief of Naval Operations." I was doodling on the bar napkin in deep black ink. I underlined the words: 'Stuff We Can Talk About.' "So what are you willing to say about starting work at Central Intelligence?"

"You referred to it as my post-retirement career. I simply called it my second career, and as I later got ready to leave the CIA and the government and undertook some consulting contracts, I set up a file in my filing cabinet which I labeled "A Third Career" and pursued that with some minor details for awhile."

"I thought the Third Career was helping your wife Billie when she got sick?"

Mac smiled ruefully. "That was actually a fourth career and the most important one. It opened my eyes to a lot of things. Plus,

I got my diagnosis of prostate cancer in all that personal chaos and wound up as a mentor and volunteer for the support groups for people who were just starting down those difficult paths. I am glad that sometimes I can still have a beer."

I raised my glass in agreement. "OK—we talked about your getting recruited by Bronson Tweedy to come work for Director Helms— that was like the Monday after you retired from the Navy, right?"

"Yes. I was on active duty in DIA as chief of staff in 1971, and I was considered during the later months of 1971 as a candidate to become Deputy Director of DIA. That would have gotten me a third star and increased responsibilities. I won't go into detail on how that procedure worked. I think it's generally known how the Joint Chiefs of Staff make their selection for those jobs, and then there are interviews held throughout the

Pentagon with certain officials. Recommendations are made, and finally a selection is made. Even to those of us involved, I sense that it is a very ethereal process, and I'm not sure I understand completely even yet. But I was nevertheless a candidate to become deputy director of DIA, and the only interview that I went through was held by the Director, Defense Research and Engineering, or DDR&E."

"I am glad that I had 28 years speaking in acronyms. I can actually understand what you are saying without the explanation." I took a sip of wine to help me focus. Actually, I wasn't quite sure what any of that alphabet soups meant any more. "Who was it?"

"I can't recall who DDR&E himself was at that time, but one of his principal deputies was a gentleman named Gardner Tucker who was on detail to DDR&E from IBM."

"I worked for IBM for a surreal year a decade ago," I said. "I liked them. And I liked the Big Blue motto: THINK."

"Trust me, we did. I had come to know Gardner Tucker rather well, because I was a member of the vendor selection committee for the Joint Chiefs of Staff computer program in a prior year or so, and Gardner Tucker, as a computer expert from IBM, was one of the key participants in that effort. It was when JCS was trying to select a computer program to support its command and control system: WWMCCS. I was a member representing DIA on the WWMCCS Vendor Selection Committee."

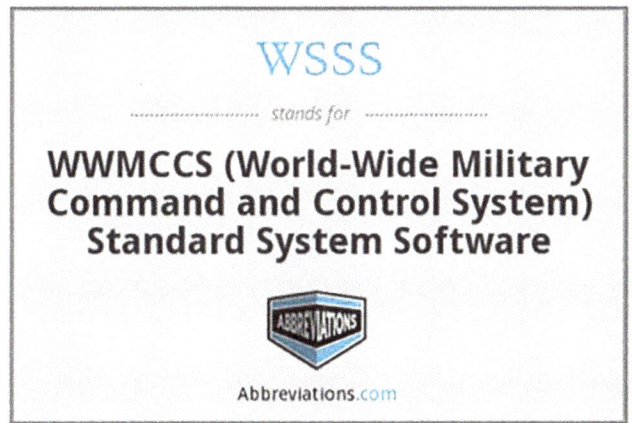

A TITLE SLIDE FROM ONE OF THE PENTAGON POWERPOINT RANGERS DETAILING THE ACRONYMS OF THE COMING DIGITAL AGE, THIS ONE BEING WWMCCS-SSS.

"I am going to bet that was the Worldwide Military Command and Control System. I think it was pronounced "Whim-Icks." It was supposed to connect all military units worldwide in one network. What year would that have been?

"Probably 1968 or 1969. I think it was while I was Assistant Chief-of-Staff for Plans and Programs for DIA rather than Chief-of-Staff, which I was during 1970 and 1971. But I was interviewed by DDR&E,· and one of the questions they asked me, was "If you were deputy director of DIA and a dispute should ever arise, or an issue should ever arise between the Secretary of Defense and the JCS (the OSD Staff on the one hand and the Joint Chiefs of Staff on the other hand), which faction would you likely favor?" I promptly responded "the Joint Chiefs of Staff."

"I completely agree. But that is sort of honest for Washington, isn't it?"

Mac laughed. "Yes, but my reason for doing so was because the Joint Chiefs, in my view, were responsible for the operating military forces of the United States. And if an issue should arise they would need direct and timely support. As I recall, the interview was terminated very shortly after that, and I was no longer, to my knowledge, a candidate for Deputy Director of DIA. I believe the DDR&E authorities were sufficiently disenchanted with my response to that question. I fell out of contention very rapidly. Another element true at the time was that Vice Admiral Noel Gayler from the Navy was Director of NSA; Admiral Harold Bowen was in the Assistant Secretary of Defense for Intelligence office, and Admiral Rufe Taylor was at CIA as the Deputy Director of Central Intelligence. The Air Force had suddenly dug in its heels, saying, "All these three star intelligence billets are being given to the Navy, and we will be *damned* if they will get the deputy directorship of DIA as well."

"The Air Force invested a lot of resources in standing up DIA at the beginning, so I am not surprised by their reaction. I think they thought of the joint activities as their natural habitat."

"Possibly. So, I may have lost out on purely inter-service political grounds because the gentleman who was appointed deputy director at that time was Lieutenant General

LT. GEN. JAMIE PHILIPOTT, USAF. PHOTO: DoD

Jamie Philpott, USAF. He became General Bennett's deputy, and I became the Chief-of-Staff."

"Was he any good?" I asked.

"No. A waste of space, in my opinion. He used to have an impeccably clean desk that he would put his feet on. When the Director was traveling, you could be sure no decisions would be made until he got back. I tell that story because that pretty much marked the end of any reasonable possibility in that time frame for me get a third star and advance into further participation in intelligence work, national, naval, or whatever. Instead, I became DIA Chief-of-Staff, and I had served for nearly two years in that position."

"That job had a lot of influence as the third senior in the Agency."

Mac nodded. "Then, in the fall of 1971, what was known as the Schlesinger Report came out. It was named after James Schlesinger who had chaired a study group on reform to the clandestine services. He recommended major changes to the intelligence community."

"Well hush my mouth," I said, "and stop me if I have heard this before," I laughed. "We are always moving the deckchairs around after the boat already hit the iceberg." Mac smiled knowingly, and I thought about having another glass of wine. Mac's story was really about creating the world that the rest of us live in.

GOING ASHORE

• • •

June, 2011. It was a delightful afternoon at the Amen Corner of the Willow Bar. The usual suspects were drifting in and Old Jim anchored the bar at his customary stool at the apex. Mac had driven his champagne-colored Jaguar sedan across the street from his apartment in The Madison, the assisted-living high-rise across Fairfax Drive.

We could have talked about Midway since it was the anniversary—the 69th—of the Battle that made Mac's team of code-breakers into legends in the Intelligence Community. But we had pretty much beat his war time stories to death and I wanted to hear about what came after Mac's active duty time. I had a stack of pristine cocktail napkins and a pen. A glass of Willow's Happy Hour White was near my right hand, and a thirst for both stories and some life-giving alcohol.

Mac was in his usual aloha shirt topped with a sports jacket, looking quite dapper as usual. He was on the wagon again due to the drugs the oncologist had prescribed him for the prostate thing and his heart, and Willow tried to make up for it by making his Virgin Mary into a veritable cornucopia of jumbo olives, celery, cherry tomatoes, and a pickle spear.

CIA DIRECTOR RICHARD HELMS. PHOTO: CIA

"That looks like a salad more than a glass of tomato juice," I said.

"Mac smiled. "I won't have to have dinner at The Madison tonight, that is for sure," he said with a chuckle.

"It is time to hear about your life after the Navy after Bud Zumwalt made you go ashore and retire. We agreed we are not going to talk about the details of your second career, right?"

"Mostly. We can talk about the Foreign Intelligence Surveillance Court and things like that, but there are some matters that are still a little sensitive after all these years."

"Like Project JENNIFER?" I said, using the code word people know about from the press and not the *real* name of the program.

Mac would not bite and shook his head. "Nope, not going to go there. But I can talk about Bronson Tweedy. He was old school, and the strong right arm to Director Helms."

"Helms was Director for longer than most, wasn't he?"

"Yes. Mid-sixties right up to 1973. He was old school. He had been Naval Intelligence in New York City, working on the Eastern Sea Frontier plotting U-boats when a friend approached him to join the OSS's Morale Operations Branch. They did the black propaganda. He was a Spook the rest of his life."

"It is interesting that the Navy Reservists in New York were in the middle of everything, isn't it?"

Mac smiled. "They were their own Navy, that is for sure. They ran the Lucky Luciano connection with the Mob to keep the docks safe from Axis saboteurs."

"In real life a lot of them were prosecutors and cops and stuff, right?"

"It was all mixed up together: military, law enforcement, and justice. It was actually sort of a parallel universe."

"In addition to the usual counter-intelligence work, they ran the scientific exploitation of the former Nazi scientists out on Long Island after the war."

"Yes, the projects that came out of the Castle were of extraordinary value to CNO Arleigh Burke, who was creating the Nuclear Navy."

"But you went to work at F Street at the IC Staff?"

"Not at first, and that wasn't the name. I think we were in the Original Headquarters Building at Langley. Long before there was a new one."

CNO ADMIRAL ELMO "BUD" ZUMWALT. PHOTO: USN

He took a sip of his Virgin Mary and seemed to be concentrating on something far away. "I was still on active duty in the fall of 1969. Bud Zumwalt was on a tear to get every admiral who had been senior to him to retire."

"I have heard about Zumwalt and his Z-grams and whiz kids. He really shook things up in the Navy after he was put in charge."

"I didn't want to go ashore, but the CNO wanted my number to promote my friend Rex to flag rank and that is just the way it was.

There was no animosity between us; Rex was Zumwalt's guy from his days at NAVFORV (Naval Forces Vietnam), and that is who he wanted to be the Director of Naval Intelligence (DNI)."

"It was in October 1969 that I was approached by Bronson Tweedy, who was Helm's Deputy and a career Spook very much like him. He was born in London to American parents; he went to school there in the 1930s, and lived with a family in Germany to get acquainted with their customs, language and culture. He arrived to start his visit the day Adolf Hitler became chancellor. He was a Princeton guy with a degree in European history, and went into the advertising game at Benton and Bowles on Madison Avenue before the war. In 1942, he volunteered for Naval Intelligence and served in North Africa and Europe interrogating captured German U-boat crews."

"Naval Intelligence again," I said in wonder. "Was he part of that secret POW camp the Army ran down at Fort Hunt?"

Mac nodded. "I am not sure, but it would have been Bronson down to the ground. That was the interrogation program code-named 'P.O. Box 1142' used against prisoners who were deemed 'high value,' like U-boat commanders."

"I have walked around the grounds where the camp was located. You would not have known anything happened there, since they tore everything down right after the war except the old Coastal Artillery batteries. It is an interesting story, and it was also probably a violation of the Geneva Convention. You know, there are still things that people don't want to talk about that went on at the intersection of operations and maritime intelligence."

"I agree. I touched a live wire one time from the early 1950s," I said. "I was working on a story about a counterfeit ring in France in the 1950s a while back with Tom "Big Smoke" Duvall and touched a live wire. Tom told me to back off and tell the story the way he wanted it or drop it. It might have had something to do with the intelligence connection to Lucky Luciano and the Mob, but I don't know for sure and I was smart enough not to ask."

Mac nodded. "After the war, Bronson briefly returned to advertising before being recruited by the CIA. He served in Switzerland, and DC just as the Agency was being formed, and he was Chief of Station in Vienna and twice in London. Then he founded the Africa Division, which was a result of Eisenhower's dislike for Patrice Lumumba."

CONGOLESE LEADER PATRICE LAMUMBA. PHOTO: AP

"Did he have anything to do with the coup and Lumumba's death? I remember the revelations about the rubber gloves and lethal toothpaste they were going to slip into the Congolese President's bathroom. It was as cool an idea as the poisoned cigars they were going to try to get Castro to smoke."

"I assume that is the case, but I didn't have anything to do with it personally. And the toothpaste ploy makes sense. Lumumba did have a brilliant smile, from what I recall. He died right before John Kennedy was inaugurated, and Bronson was in Leopoldville around that time, but we never talked about the list of shady things that later came to be known as being part of the Agency's Crown Jewels. After that, he was tapped to head the Eastern European Division. When Dick Helms was confirmed as Director in 1966, Bronson moved up to be Deputy."

"There was something going on in those years," I said. "I mean, someone got away with killing the President of the United States. The Warren Commission had so many glaring flaws that everyone suspected it was not the open and shut lone gunman that the Report claimed it was."

"We have talked about that before," he said, looking around to see if anyone seemed particularly interested in the topic. "I think there was a Texas connection. When Nixon came into office in 1968, his people immediately focused on the Intelligence Community. Henry Kissinger thought he had all the answers and viewed the Agency with condescension."

"He still does, from what I hear."

"He thought the boys from Langley were not sharing all that they knew with the Administration. Nixon felt that steadily increasing capabilities and costs directed toward IC functions should be yielding better analysis. Plus, the coup in

Cambodia in '70 caught everyone by surprise, and Nixon hated being surprised."

"Since there did not appear to be a direct link between level of effort, and the money spent to produce it, Nixon commissioned James Schlesinger to conduct a survey of the IC. His chartered goal was to identify problems within the IC and recommend ideas for improvement."

JAMES SCHLESINGER AS SECDEF, 1970. PHOTO: DoD

"I have been to that movie before. I think they are doing it again now," I laughed. "So all that is swirling around while I figure out what to do as I am being I am being pushed out of the Navy. Bronson must have heard about it on the grapevine. He gave me a call in October of 1969, and asked me to come down to the City Tavern Club on M Street and talk about a proposition he had for me."

Prep work for the dinner trade complete, Willow Co-owner Tracy O'Grady came out to press the flesh with the usual suspects before the kitchen got busy with the dinner trade. She worked the Amen Corner there by the front window for a while as Old Jim showed her his latest flight of blank verse from the notebook he kept in his corduroy jacket. Jim is still quite the poet. My pen was still poised. "Ok, Sir, you retired in 1971?"

"Yes. November of 1971. Bronson Tweedy called me and asked if I could meet him for lunch at the City Tavern Club on M Street in Georgetown, which I agreed to do. He liked that place-very traditional Washington institution. At lunch he said, "In view of the Schlesinger study making these demands on what the newly established Director of Central Intelligence is supposed to do, the DCI has decided reluctantly that he will have to expand his budget staff to carry out more coordination of the intelligence community. And among other things we'd like you to come and work for us."

"I asked him at that point if he wanted me to come in uniform or if he wanted me to come as a civilian. He said, "We want you as a civilian.""

CIA DEPUTY BRONSON TWEEDY

"When was that, Admiral?"

"This was probably the first week of November. Bronson said: "We'd like to have you on board by the first of December." I told him that I doubted that I could get "unhooked" from the Navy that quickly, but I would try. After a pleasant lunch, I returned to the Pentagon and made inquiries, and the first thing I was confronted with was that General Bennett, the Director of DIA, was away on a trip. I was told that there was no way that I could send my request for retirement without his endorsement and agreement, which was obvious. So, it being readily apparent that I couldn't carry get on the retired list by the first of December, the earliest I would be able to do it would be the first of January the following year. That was agreed to, and that's what happened. As soon as General Bennett returned from his trip, I had my letter on his desk requesting retirement. He endorsed it, I went through the necessary procedures, and I was retired as of the 31st of December 1971 and went on the retired list on the 1st of January 1972."

"I was still worried about the Draft then," I said with a sigh. "I was dodging the draft and hoping I wouldn't get nailed as soon as I graduated. Did you have any regrets about going to the CIA after all those years in Naval Intelligence?"

"No. It was a good offer. I don't remember the pay scale at the time, but Bronson Tweedy's offer to me was that, "We will take you on as a contract employee. We'll give you a one-year contract renewable. And we will pay you the equivalent of a GS-16 salary."

"That would be a General Schedule employee equivalent to a Rear Admiral, right?"

"Yes. The concept of the Senior Executive Service did not exist then. Compensation was about the same as what I was making from the Navy. I knew that I would have to forego part of my retired pay. I think the formula at the time was that I'd have to lose half of my retired Navy pay while I was in government employ and have that restored to the full annuity upon leaving government service.

But I would concurrently be getting a full civil service salary or salary from the DCI, which would really give me a pay-and-a-half and make me a real true "double dipper," a status for which I was accused of many times."

"It always seems to irritate some people around here when somebody in the military finally gets a decent salary. With full military retirement and a job, you can actually afford to live in DC. At least you did not have to go into bid-and-proposal work with the rest of us Beltway Bandits."

"I am thankful for that," he said, taking a nibble of the celery stalk in his Virgin Mary. "New Year's Day of that year fell on a Sunday, so we had Monday off to observe the holiday. I believe I retired on a Friday, and went to work at CIA headquarters on Tuesday. I know I had a three-day break between careers Time enough to have a New Year's party and recover from it."

"That only means you were not trying hard enough," I said with a snort. "You told me about the party the senior officers had at Joe Rochefort's house during the War after the word came back that Station HYPO had been right, and the Japanese were shattered at the Battle of Midway."

Mac nodded solemnly. "We did not see some of them for a few days. But this was no war and we were all a little older. When I arrived at the DCI headquarters, I first went into a group that was headed by J.J. Hitchcock, who was one of my previous friends in naval service."

"More Navy" I said, underlining my notes.

"Yes. I had first met J.J. at the Naval Security Station back in

'47-48 when he was doing some research work on indications and warning. J.J. had become the I&W expert for the DCI over the years. He was instrumental in setting up the Watch Committee and doing the Weekly Review of worldwide Indicators and he issued the weekly Watch Report that was a major instrument of power in the government during those years. By then, though, J.J. wasn't doing that kind of work any longer. He was simply doing staff work."

"Did you work on the 6th Floor at Langley? That is where we had our offices on the Community Management Staff after they changed the name again."

Mac looked contemplative. "Could have been. That seems right."

"That is the subject of the first major review of the way the Intelligence Community worked. There was some thought that once the war in Vietnam was transferred back to the Republic of Vietnam, there would be plenty of budget authority to transfer to other more strategic tasking."

"I have heard about the Williamsburg Conference when the big post-Vietnam drawdown was going on. The military divided up all the responsibilities for the DoD components of the IC."

"That came later," said Mac. "Once we had pretty much extricated our forces in the field in 1973. The Schlesinger report landed on our desks for implementation, with some lofty observations. It claimed the line between 'military' and 'non-military' intelligence had faded; scientific and technical intelligence with both civilian and military applications had become the main battery for the community. All the other stuff was sorted according to the people that used it. The strategic stuff was for the national decision-makers, and the tactical stuff they didn't care about was for the regional and functional Joint Commanders."

"Like CINCPAC and the strategic Air Command?"

"Precisely. The President and the National Security Council with Kissinger were served mostly by CIA for the national-level stuff, though NSA was the critical collector for special SIGINT. But there was more, and it was urgent. The rate at which the Soviets were cranking out innovative technology revolutionized the intelligence cycle."

"We called it Tasking, Production, Exploitation and Dissemination," I said. "TPED for short. I don't know what they call it now. Find, Fix, Finish, Exploit, Analyze, and Disseminate?"

"I can't keep up with it," said Mac with a lopsided smile. "That is the same concept we used starting with Operation intelligence in the Pacific War. Nothing changes except the acronyms."

"So that is the 1970s under Richard Nixon at Langley."

"Yep. Before we moved downtown and the two Congressional Panels blew the bottom out of everything. That is worth a conversation all on its own."

I underlined a couple of Mac's quotes on the square white cocktail napkin in front of me, and added the Congressional Pike and Church Commissions to my notes. "I will do my research, Sir, and be prepared to discuss them when you feel up to it."

"I am always ready," said Mac. "Those were some interesting times."

I smiled and waved at Big Jim the bartender for the check. "Isn't that a Chinese curse, Admiral?"

"Only if you are uninterested," he said with a grin.

PART THREE:
AGENCY BUSINESS

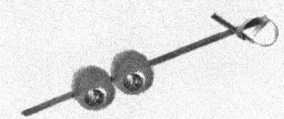

A MATTER OF TRUST

• • •

TINKER BELLE HOLDS DOWN THE BUSINESS SIDE OF THE WILLOW BAR.

Willow was having one of those golden afternoons where the doors to the patio were propped open, and the dust motes glowed in the rays of the lowering sun.

Jasper was running around as always—I don't know when he *doesn't* work—and Tinker Belle was handling the mixology behind the bar, awaiting Brett's arrival at cocktail hour to handle his second job of the day. He is actually a computer programmer, or a movie star or something most of the day.

Mac had motored over to the bar in the Jag, claiming to feel a little tired, but he was ready to speak, and I let him. Old Jim was listening to his MP3 player at the apex of the Amen Corner.

"So, it is the mid-1970s. You are moving around in the various staff elements of the IC Staff as a special assistant.

"I was. There were other personnel who were coming and going, as there was considerable turbulence in the staff, both personnel-wise and organizational. It was natural since the Pike and Church commissions had made so many demands on the IC. As the DCI's agents for compliance, we tried to feel our way and better establish ourselves in the Community. We were still housed in the CIA building at Langley. Later, we moved down to the old Selective Service building near the White House since everyone thought that because we worked on the sixth floor of the Original Headquarters, we were a wholly-owned subsidiary of the Agency."

"The Agency still thought so, too, and that was twenty years later."

Mac smiled over the rim of his glass. "We were still treated with suspicion by most elements of the IC. The work that we were doing, except in the program and budget areas, continued to be highly suspect by all concerned. Danny Graham took a crack at being the director of the staff, I think temporarily, before he became Director of DIA. He had served with us on the staff from the Schlesinger days, and he moved up to be interim director before Sam Wilson came. But I think it's safe to say that, after Bill Colby got himself well settled in as DCI— he having been an old hand in intelligence, and he clearly knew where he was going—when Sam Wilson came in as Director of the IC Staff, I think Sam more than anyone else put order into the work of the IC Staff. It wasn't something that he could do too quickly, and it evolved over time. But he was there long enough to do it, to actually accomplish it."

"There is an ebb and flow to staff organizations. Look at the Director of National Staff. It is getting huge since it was implemented. But you were done with Program and Budget and were a trouble-shooter for Director Colby?"

DCI WILLIAM COLBY TESTIFIES TO CONGRESS.

Mac nodded. "The DCI performed functionally and legally as the *Pater Familia* of intelligence. In a sense, we were clearly a projection of the DCI because he had legal responsibility reinforced by subsequent congressional actions to oversee and coordinate the whole shooting match—responsibility without the complete authority to tell people what to do. But many departments and agencies in the community outside of CIA continued to treat us with circumspection, if not outright contempt. They never took us fully into their confidence on things that they didn't think we needed to share. We were effective in coordinating and consolidating the program and budget."

He took a sip of beer and smiled in pleasure. "In other activities," he continued, "We didn't fare as well. In order to gain respect, and I think that's the best way to word it, the IC Staff, as it grew, tried to do things that they conceived as perhaps not being done well or as completely by the individual elements of the community and that could maybe be done better and more effectively centrally. That's one reason that we got into post-mortems, for example."

"What was a 'post-mortem?'" I asked. Something like the USS *Pueblo* Damage Assessment you worked on at DIA?"

"Yes, but more comprehensive. The thinking was that the IC Staff should look over the involvement of several agencies in a given action from a strategic perspective, assess if it could have been done more effectively, and make recommendations on changing things."

"I imagine that would be quite a threat to the agencies if they thought they were being second-guessed after the fact."

"You bet. I think the post-mortem effort fell on its face because the other agencies, if they had performed poorly, didn't want to share it with the IC Staff or other elements of the Community who would try to grab their resources or mission. We did try, though. I participated in the *Mayaguez* post-mortem."

"That was the official last combat mission of the war in Southeast Asia, and we left Marines behind."

"Yes, and not a pretty tale. But I knew we were not getting the full story from all of the agencies and I *knew* we were not going to be able to write reports that would be complete and accurate. But as usual here in Washington, nothing is either of those things. I was happy in 1975. At that time, certain legal actions of the Attorney General put a job on the DCI's that required full-time attention and the continuity of at least one person in order to make sure things went smoothly. All I can say was that it was in the electronic surveillance area."

"You are being cagey, Admiral. Is that what we are *not* going to talk about?"

"The very one," he said, taking a sip of his beer. "I worked that program until I retired for the second time. Those duties took up about half of my time. In the rest of it, I worked things transition issues between the Ford and Carter Administrations."

"Now you getting into territory I actually remember," he said with a laugh.

JFK AND TED SORENSON AT THE WHITE HOUSE. PHOTO: AP

"Whippersnapper," he said with a laugh. "I helped brief Ted Sorensen prior to his confirmation hearings, and frankly it was no surprise to me he didn't get confirmed. I thought in briefing him that he had relatively low regard or low appreciation for security. If you look it up when you get home, I think you'll find he was charged with some security laxity over time that didn't speak well for himself and for his earlier service to President Kennedy."

"Admiral Stansfield Turner became second choice, and he was brought in as a nominee. Admiral Turner spent, I think, approximately a month getting ready for his confirmation hearings. He was given an office in the Old Executive Office Building with a small group of people, most of whom he brought in with him from duty stations in the Navy. Some of the people had been with him since he was at the Naval War College at Newport and also had been with him over in Naples. They were neither intelligence officers nor intelligence experts, but the point was that Admiral Turner clearly relied on them ahead of the advice he got from other people."

I smiled since it was a familiar story. "I knew one of them—I forget his name—but he was on the Senate Intelligence Committee, and he told me once that it was eerie working there. He said the Turner people were treated as almost being invisible."

"I think that is an accurate assessment. Of course, we were downtown and out of the Headquarters by then, and out of sight, out of mind."

"So you had an Intelligence Community that did not trust the IC Staff and a DCI who trusted no one."

"That is about the size of it. Our time with Admiral Turner was not a placid one."

"But you kept working right up through the election of Ronald Reagan as President?"

"I did. I was a special assistant to Bill Casey. That was an interesting man."

"I bet. And then you retired again in 1984?"

"Yes. We discovered that Billie was ill with something really sinister. Alzheimer's, the early onset kind. It was time for me to stay home and become a caregiver."

I gave a little shiver since Dad had just succumbed to that cursed disease the past January. Watching his steady decline from dashing Naval Aviator and yachtsman to wasted shell was a wrench for five long years. "I know how awful the process is. You helped me a lot in knowing how to deal with it."

"We managed to keep Billie at home for almost ten years. And that is what got me involved with volunteering with the support groups. But that is another story."

I nodded in agreement. I reached for my wallet after underlining the phrase "fucking dementia" on the cocktail napkin in front of me. In many regards, Mac's third career might have been the hardest of them.

SEMPER FI

• • •

TIM, MAC, AND JOHN EARLY THIS YEAR.

"If the Army and the Navy Ever look on Heaven's Scenes
They will find the streets well-guarded by The United States Marines."

The word was passed later last week. Lt. Col. John J. Guenther, USMC (Ret), 79, passed away peacefully at his residence at The Jefferson in Arlington, Virginia, on 29 October 2009. A mutual friend, a colleague, and a neighbor reported that John had been in declining health for the past few months. He had just observed his 79th birthday.

There is going to be a Mass was held in his memory at St. Agnes Catholic Church in Arlington, VA, on Wednesday, November

4. A funeral Mass at the Ft. Myer Chapel and funeral with full military honors will be conducted at Arlington National Cemetery on Thursday, January 14, 2010, followed by a reception at the O' Club. The death notice for John appeared in the Washington Post this morning. I scanned it but there was not much in it, except the personal information about his family.

John's family was a lot larger than just those of his blood. In fact, it is more than a Corps.

Mike Decker, who relieved John at USMC HQ, commented on a long and distinguished career this way: "John Guenther enlisted in the Marine Corps in January 1948. In September 1950, Corporal Guenther landed at Inchon with the Seventh Marines. In November and December, Sergeant Guenther was in the S2 during the Chosin Reservoir battle."

THIS PHOTO IS FROM THE FIRST WEEK OF NOVEMBER 1950, DURING THE BATTLE WITH CHINESE FORCES AT SUDONG-NI. SERGEANT JOHN GUENTHER IS ON THE LEFT.

"He served other challenging/tough tours in Cuba, Vietnam, and East Germany. He was a great Marine who embodied the full measure and meaning of Semper Fidelis. John served more than thirty years in the Marine Corps and, upon retirement, continued to serve in a civilian capacity, rising to the civilian equivalent of BGen and serving as the Assistant Director of Marine Corps Intelligence."

That is where I first met John in the mid-1980s, in the old threadbare Navy Annex on the bluff above the Pentagon. He was a legend even then. Based on his contributions to the Marine Corps across six decades, which included active duty and service as the senior USMC civilian, he is widely credited as the father of modern Marine Intelligence.

On hearing the sad news, Joe Maz commented: "John Guenther was a friend and mentor to many of us. Most importantly to me, though, he was a living connection for my WWII Marine Corp Dad who fought at Guadalcanal, Cape Gloucester, and Pelielu who I lost in 1996. There are all too few of our WWII heroes left. Semper Fi."

I had a chance to work with John early this year on an article he did on Strategic Surprise during the North Korean and Chinese invasions of the South. He described his reaction at seeing his first dead North Koreans hanging from an old Soviet tank and just how cold it was walking out of Frozen Chosin.

The article was part of something larger John wanted to complete before his tour was over. Mac says that recently completed and submitted to Marine Corps Headquarters a thorough history of Marine Corps intelligence, a project that he has worked on for the past several years.

All of us have lost a very good friend, Marine or not. He will be deeply missed. Semper Fideles, John.

Always Faithful.

MAC'S PORCUPINE BALLS

● ● ●

A groundswell of interest rose over Mac's revelation at Willow that he is a past master of the pressure cooker, the most dangerous single device in the Socotra Test Kitchen. Check that. The pressure cooker is the only device that actually could produce high-velocity shrapnel during the cooking process.

I should interject here that our friendship began over common interest, and the totally logical decisions Mac did. His wife Billie was a fairly high-powered real estate agent in the northern Virginia go-go property market. She had the duty at the office one weekend and realized that she could not operate the PBX system that she once operated with ease.

She also realized she was getting lost in places that she knew well. It was the beginning of something that was impossible to imagine: the theft of her memories and eventually her consciousness.

Mac did the right thing, as he always did in his life. He became a caregiver, and a coach for others who were entering the bizarre world of the "36-Hour Day" that families experience with the early onset of dementia.

Mac was able to keep her at the pleasant split-level home in North Arlington for a decade before it became too much, and Billie needed the care of a dedicated nursing staff. Along the way, Mac gave himself to mentoring support groups for both the Alzheimer's community and Prostate Cancer support groups that were sponsored by the Virginia Medical Center on George Mason Drive here in town.

There were also children and grandchildren to be fed, and so Mac dedicated himself to learning how to cook. We were at Willow one afternoon—I was in a foodie phase of my life, and I was interested. I was putting together a compendium of recipes from Spooks who had representational obligations at home and overseas, the can't miss ones that work even on short notice. I asked him about his favorite recipes. He took a sip of his ginger ale and thought for a moment.

"I liked the pressure cooker. It saved a lot of time. The deep-fat fryer is fine, and also provides a rich threat environment for the sensory altered experimental chef, and that is not at all a reach to the most deadly industrial cooking device, the pressure fryer. That device is normally found in fast-food restaurants that specialize in chicken, and the beauty of it is that is combines the inherent danger of really hot oil with the possibility of catastrophic explosion."

The is, by the by, precisely how an itinerant gas station owner from Corbin, KY, got his start as a global icon. Harland Sander's pumped gas and sold fried chicken, which he served in his house next to the gas station. That might have been that, save for the expansion of the Eisenhower Interstate System in the 1950s. I-75, the fabled pavement that generally follows the path of the old Dixie Highway from the Soo Locks to Alligator Alley in Florida, by-passed once bustling Corbin, and at the age when most of us are dreaming of social security, Harlan found his gas station out of business.

KFC ICON COLONEL HARLAND SANDERS IN HIS TRADEMARK WHITE SUIT. IT'S RUMORED THAT THE ORIGINAL WAS FASHIONED WITH AN ARMOR PLATE DUE TO THE RISK OF EXPLODING PULLETS.

The Colonel took the proceeds from his first check from the SSA and started driving around middle America with a secret mixture of a few dozen spices and a potentially deadly pressure fryer. The strangely compelling chicken he produced changed the history of Japan, among other nations, and led directly to *Loving That Chicken From Popeye's* and the outbreak of obesity in neighborhoods adjacent to the franchises.

I would like to experiment with an industrial plant but have not got around to equipping the Test Kitchen with a rig. But I digress.

Mac said he would bring his old index card with his favorite recipe to our next meeting. "I can't cook at The Madison anyway. You can have it."

Next time we met, I looked at the card, titled "Porcupine Balls."

"That is sort of unsettling," I said.

"It is a tasty mealtime treat that is fast and easy," Mac said and proceeded to walk me through it.

"I don't have a pressure cooker," I said. "Too scary."

"Not a problem," replied Mac. "You can do this one without resorting to National Technical Means. Let me explain it."

He was a little foggy on exactly how long he cooked his meatballs in the pressure cooker, and I expect it is a trial-and-error thing, waiting either for the explosion or the perfect duration of the food under pressure. You can't just open the lid during the process to see how you are doing, after all, or rather, you can but will probably only do it once.

For a guy who won a couple of global struggles, I accept that. But here is a higher and safer road for Mac's Balls:

Mac's Porcupine Balls

- 1/2 cup uncooked long grain rice
- 1/2 cup water
- 1/3 cup chopped onion
- 1 teaspoon salt
- 1/2 teaspoon celery salt
- 1/8 teaspoon pepper
- 1/8 teaspoon garlic powder
- 1 pound ground beef
- 2 tablespoons canola oil
- 1 can (15 ounces) tomato sauce
- 1 cup water
- 2 tablespoons brown sugar
- 2 teaspoons Worcestershire sauce

DIRECTIONS: IN A BOWL, combine the first seven ingredients. Add beef and mix well. Shape into 1-1/2-in. balls. In a large skillet, brown meatballs in oil; drain. Combine tomato sauce, water, brown sugar and Worcestershire sauce; pour over meatballs. Reduce heat; cover and simmer for 1 hour.

Yield: 4-6 servings.

NOTE: This is a guaranteed non-explosive dish. Really.

POSTHUMOUS

• • •

THEN-LT JOE ROCHEFORT. A GREAT AMERICAN, LIKE MAC.

I have to tell you, Mac's bottle of chardonnay was not a modest vintage. It had woody notes and was inviting in aroma and crisp in presentation. So was President Reagan when he gave the Distinguished Service Medal to Joe Rochefort.

The years fell away as I looked over at the binder on the coffee table in front of Mac. The first picture had the iconic movie actor whose greatest role was that of President. He looked fabulous, just as I remembered him from before the time his mind was stolen like Mac's beloved Billie, and the once-entrancing gaze became glassy and the jowls drooped regardless of what Mac tried to do.

"When was this, Admiral?" I asked.

Mac looked down at the binder. "May 30th, 1986," he said. "I had been working on it for three years, the first while I was still working for the DCI. Jasper Holmes tried, in the mid-1950s, when he was back to teaching at the University of Hawaii. He got shot down, along with Fleet Admiral Nimitz. The Redman brothers and that pompous Wenger character were still around at NSA and in Naval Communications. They awarded *themselves* the DSM and could not bring themselves to admit that they assassinated the character of a better officer than them."

"Joe Rochefort's leadership at Station Hypo was one of the key aspects of the victory at Midway. And the way he was treated was shameful and embarrassing," I said.

"True. But when President Reagan gave the DSM to Joe's family, there was justice. Of course, only a handful of people who were there knew it, but of course, nothing is easy. It gets stranger. Joe's son and daughter were not talking. They had to have two medals and two citations."

"Families are funny," I said, thinking of my own. I looked curiously at the picture of the two Rochefort kids with the President of the United States, no animosity apparent.

"The son was an Army Colonel," said Mac. "Joe Junior. I guess he didn't want to get into a whole career fighting his dad's battles. Joe Senior died in 1976, but he remembered the insult to his last days. His daughter—Janet Elerding—she was there, too, with her whole family."

"I knew you took on the fight to get the medal awarded even though the statute of limitations for the award had passed."

Mac nodded as he flipped through the Kodachrome 8-by-10s.

"I got through to Jasper Holmes, who was still alive then. He was in a nursing home in Honolulu, and it took a while to get him to understand what we had pulled off. But when he realized what I was talking about, he said it was Justice, finally. There was quite a tale about how it all came to pass."

Distinguished Service Medal. It is second on the precedence list of Navy awards, following only the Navy Cross.

"So the medal was denied twice? I didn't realize that."

"Yep. Fleet Admiral Nimitz recommended it to Ernie King, but on the advice of the Redmans, his chief of staff denied it. Then, it was turned down again after the war when Jasper tried to push it through. Bu the time I started, the three son-of-a-guns who torpedoed his cryptologic career were all in their graves."

"SECNAV was John Lehman then?" I tried to remember. Mac nodded affirmatively.

"Yes. The package got lost on the OpNav staff, and it took a while to get it in front of him. But he was a believer."

"Confusion in OpNav?" I snorted. "Go figure. But the second

Reagan administration was a good one for the Navy. Remember the five hundred ships and Strategic Home Porting?"

"Yes, I do," said Mac. "I don't think it was just Star Wars that sunk the Soviet Union. The whole thing is very much a Cold War story, but the Iranians are in there, too."

"What do you mean, Admiral?"

"Well, I was working for Bill Casey before I retired from the Community Management Staff in 1981. I mentioned to him that we should try for an IC award since the DSM had been turned down twice. Casey told me that it was a Navy victory and it ought to be a Navy medal, and we decided to go for it. The Iran Contra affairs had given the IC another black eye . . ."

"Like when *don't* we have a black eye? If we have a success, we can't talk about it, and if we have a failure of any kind—operational or intelligence—it is always an Intel fuck-up."

Mac laughed. "That has always been true, and even triumph only got Joe Rochefort transfer to command a floating dry-dock after making the primary contribution to the biggest naval battle since Trafalgar."

"So this was all sort of a perfect storm that got Joe his medal?"

"It was. The package we put together never got to John Lehman. It had not been rejected as we thought. He didn't know about it until Eddie Layton's book was coming out, and there was a lot of publicity about the communications intelligence we could finally talk about.

George H.W. Bush had been DCI, and he was sensitive to the morale in the Intel Community. The President was committed to supporting the Contras in Nicaragua, and this looked like a good opportunity to highlight a great success."

"Plus, the Redman Brothers were dead," I said.

"Yes, they were. But of course, almost all the players in the actual event, 44 years before, were gone. Tommy Dyer. Jasper Holmes. Eddie Layton. Joe himself, of course."

"Sometimes it is best to be the last man standing," I said. "That would be you, Sir. Who else was at the ceremony besides you and the family?"

Mac smiled. "Not a bad turn-out," he said, flipping to a type-script page with official stamps affixed top and bottom. "Let's see: The President spoke briefly and handed out the medals. The citation said that Rochefort's information 'served as the singular basis for Admiral Nimitz to plan his defenses, deploy his limited forces, and devise strategy to ensure U.S. Navy success in engaging the Japanese forces at Midway."

"Awesome," I said, taking a sip of the chardonnay and feeling the oaky richness slip over my palate and the fabric of the present give way to something else, Mac's spirit so immediate and powerful.

"Vice President George H.W. Bush was there, SECDEF Weinberger, DCI Casey and the Chairman of the Joint Chiefs, Bill Crowe. Bill Studeman was there, too, though I can't recall if he was Director of Naval Intelligence or NSA at that point."

"I need you to know this," he said calmly. He pointed to the typed memo with the stamps on it. "This is one of two things I value the most, though my family probably won't care. It came from a friend of mine in St. Louis, right out of Joe Rochefort's personnel file. It is the justification for the DSM. He flipped the glassine envelope over, revealing another handwritten letter." This is the one that I prize the most. It is the one that Admiral Nimitz wrote in 1958 when Jasper tried again to get the medal for Joe.

MAC AND HIS BRIDE, BILLIE, 1948. PHOTO: SHOWERS

"Whoa," I said. "That is a historic piece of paper. Good thing it didn't go up in that big fire at the warehouse. How did you get it?"

Mac smiled and looked off in the distance. "My pal in St. Louis said I could keep these until I was done with them." He closed the binder. "I am not done with them."

I toasted him with my glass of chardonnay. "I suppose you should get around to deciding where these papers should go, Sir, But not yet. Not yet."

Mac gave me one of those Sphinx-like smiles. "No," he said. "Not just yet."

THE PROFESSIONALS

• • •

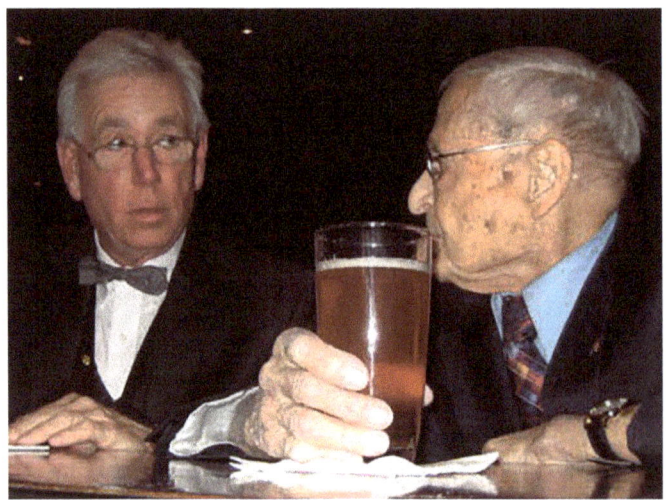

T hat night, Mac was eager to get home by eight when the President was scheduled to address the nation from the Oval Office. It was about one of the wars ending, from what I could gather, though it seemed a little premature with nearly 50,000 troops still in country.

We were at Willow with an unexpectedly large turn-out by the civilian sector. The crowd of Regulars was thin, or late in arriving is a better way to put it. Jim and Pete were holding things down, and a Marine with the right haircut and tats extending down from the sleeve of his polo shirt was holding down our usual place at the 'L' of the Willow bar.

This gathering was a little impromptu, and called with some urgency since deadlines were looming. I need to get the 25th *Anniversary* issue of the *Quarterly* on the street, and thought I would weave Mac's recollections of World War Two into a sort of tribute to both him and the organization.

I'll be completely honest with you: I have been taking notes with Mac for a long time, and it is a body of work that already exists. I would prefer to put that together rather than scour the archives for gems from the last hundred-odd issues of the newsletter. I'm lazy, I guess. But the more I thought about it, the more I liked the concept.

There are a couple of unique truths that need to be remembered.

One is about the scandalous conduct of some Americans who went to their graves with their reputations intact and did so on the backs of others who did not. And there are errors, in fact, about who should be credited with key elements of the victory that was so significant that it took sixty years to piss away the results of the sacrifice by so many. And it is The Professionals who made it possible to tell the story before it all slips away.

See, without the Organization, I never would have known Mac, and the story of how it was founded had never bubbled to the surface.

That is why Pete drove all the way up from Warrenton to get his two bits in, and I am glad he did. He was sitting out on the patio in front of Willow when I strolled over from the office. We walked in and secured the corner nook where we could stretch out and get comfortable.

I waved hello to Peter, the bartender, as we walked in and

signaled for refreshments. "Peter!" I called out. "I'll have a glass of your cheapest white!"

He looked back at me with a sly grin and nodded. Next to me, Pete considered his options. He is a beer man, like Jake, the Chairman of the Professionals, but I have given in to Willow's polished allure and its wine list.

Truth be told, the pinot grigio they serve at happy hour prices is pretty damned tasty, and a great bargain. If I started guzzling the high-test, these little sessions can start to run into serious money.

I had my pad out and a decent pen with which to scribble random factoids.

Pete's beer arrived, and Peter gave me a generous pour of the crisp white in a tulip glass. I did not know quite where to start with Pete's story, since I know him only professionally, so I imagined myself being a real journalist and decided to start at the beginning.

"So how did this all come together?" I asked. "The formal founding we are honoring this year was in 1985, and I joined in 1986. Not that I paid a lot of attention to it. Drew Simpson, my boss at the Bureau, sort of threatened all of us working in Naval Military Personnel Command (MPC) Code 4411 if we didn't sign up."

Pete leaned forward, a little distracted by the direct lurch into the interview. "Well, here is how it began. I retired out of DIA, the Soviet Navy Branch at Arlington Hall in 1977."

"Arlington Hall!" I exclaimed. "That is right across the street from Big Pink where I live. That would have been a fabulous commute!"

"Yeah," he said, probably thinking of the 50 miles of concrete lunacy that separates Warrenton from the capital. "I went to work at BDM, for RADM "Shap" Shapiro. But once I was out of the Navy, I felt like I was losing track of some great shipmates. I happened to run into Bill Armbruster at the Fall Dining In. He was the detailer at the time, and naturally had access to all the retired addresses in the database."

"I don't think they could give that information out now," I said. "All that privacy stuff."

Pete nodded gravely. "He gave me around 1,400 addresses, and I wrote a letter and had it printed, asking if anyone had interest in establishing a professional group to stay in contact. I paid for the postage and addressed them all myself."

"What was the response like?" I asked.

Pete smiled. "I got nearly 400 responses, which is pretty astonishing for a mail solicitation. We went from there."

"So who were the founding members?" I asked, trying to sip wine and keep writing legibly.

"Well, let's see. There was Tony Sesow, who later went to the Naval Intelligence Foundation side of the house.

"He still runs the annual golf tournament out in Shenandoah," I said. "I have never met him."

Pete nodded. "Yep. Tony and me, and Art Newell, of course, who never came to meetings, since he was up in Newport running his clock repair business. Art was older than us, with World War Two service. Bill Bailey had got out and was an attorney, so he was a natural pick to be on the board of the new professional organization. And Mac, naturally.

He was an Admiral, and we wanted to have a Flag officer on the board."

"I looked at the masthead on the first newsletter," I said, referring to my notes. "Wasn't Four-star Admiral Bobby Ray Inman on the board, too?"

I looked up and saw that Mac had arrived, clad nattily in his summer suit with a neatly knotted plaid tie on his blue shirt. Even with the scorching heat outside, he looked cool and relaxed.

We scooted around and got comfortable. "I heard you talking about Bobby Ray Inman, the first intelligence officer to make four stars," said Mac.

"Yeah," I responded. "What was he then? Director of Naval Intelligence?"

"No, BRI was up at CIA by then, as the Deputy Director. Bill Studeman was the DNI. They were never at Langley together," said Mac. "Of course, Bill got four stars up there as well."

"We wanted a four-star on the board," said Pete. "He agreed to do it, but he never came to the meetings. He was the one that did me in, career-wise, when I was the Assistant Chief of Staff at Third Fleet. He was the PACFLT N2 at the time, and we were taking the staff ashore and relocating to Hawaii from San Diego."

I extended my hand to Pete. "Happens to everyone, sooner or later, except for Bobby Ray and Bill. Put her there, Shipmate. I had that job, too."

"Inman says his calendar is normally full a year in advance," said Mac. "We normally had the meetings at my house in Arlington, and Art would call in from Newport."

I noticed that the Second Greatest Fighter Air Intelligence (AI) I of all time had arrived and was leaning on the bar waiting for Jim to produce a beer. He still wears his fighter crew cut, grown out a bit, and he looked stylish in a well-cut business suit. He is leaving major command to join the private sector, and I was hoping that he might join our little band of gypsies down the street. I waved at him to come over, and he almost collided with the Good Doctor, who was running late as usual.

The two men shoe-horned their way into the conversation nook where we had relocated for this potentially important meeting. , and Mac grimaced at his Virgin Mary. "I can't have olives this afternoon. Dental work today left me a little sensitive. But there is no horseradish in this thing. Anyone care for food?"

The Second Greatest Fighter AI of All Time demurred. He had a dinner with the Israeli Naval Attaché he had to get to, and shook hands all around. Mac glanced at the Neighborhood Bar Menu, which is still a great deal and Peter got the signal and came over, bringing a fresh bottle of pinot.

"Where is Sara-With-No-H?" asked Mac. The petite but fiery Lebanese waitress is a particular favorite of his, and I have to agree she is one of the prettiest women I have seen. Her dark eyes, with those exquisite lashes and delicate curved brows are deep enough to fall right into.

Peter frowned. "Sara is not with us any longer," he said delicately.

There were groans around the table. "How come?" I demanded. "What happened? This is outrageous."

Peter did not want to discuss the matter. "Let's just say there was an issue with management."

"Damn," said Mac. "I suppose we will just have to go talk to her at that other place she works. That is down at Dupont Circle, right?"

"Yep. A place called Cobalt. She works the second-floor service bar. We hung out the other night. She really resented being told that 'too cute is too hard.' She blew up and demanded an apology. In the end, she said she needed some time to cool off, and management told her to take just as much time as she wanted."

"Don't blame her, I said sadly. "We are going to miss her," I said. "How about an order of the Pork Spring Rolls, some deviled eggs, and the Miniature Fish and Chips?" Peter nodded with approval. The Good Doctor added a half order of the signature Flatbread with Shrimp, and Mac said: "This thing is going to be over by eight, right?"

I nodded. "I can't imagine going that long."

The Good Doctor looked over and said he wanted to listen to the President, too. "He is going to declare an end to the combat mission in Iraq," he said. "The United States has met its responsibility to Iraq, and it is time to turn the page and get back to the pressing problems at home."

"Is this like the flight onto the aircraft carrier and the Mission Accomplished thing?"

"No, not quite. I think the President is going to praise the troops who fought and died in Iraq, and still mention that he thought the whole thing had been a mistake in the first place."

"I understand that the enemy gets a vote in these things, too" I said darkly. "He is going to emphasize that his primary job is addressing the weak economy and other domestic issues, and I think he will make it clear that he intends to begin disengaging from the war in Afghanistan next summer."

"That is not going to be good for business," I said. "I think I have seen this movie before."

Mac cleared his throat, having seen the whole double-feature. "We used to meet at Bolling Air Force Base for years, at the NCO Club. The events of 9/11 ended *that* when they locked down the base."

"We can wrap the business end of this pretty quick," I said, looking at my notes. "I just need the quick story of how this professional organization started, and how the Foundation came to be. Was it just a club for the old Flag officers?"

Pete shook his head in disagreement. "No way. Admiral "Shap" Shapiro was always interested, but in the background. He did not assume the Chairmanship of the Foundation until we got the legal opinion that we could not distribute funds through the Professional Association side of the house. That was about two years after we founded NIP."

"Was that the original name?"

"No," said Pete. "Originally, we were just going to be an organization of retired officers, instead of a comprehensive professional association of officer, enlisted and civilian intelligence specialists, active and retirees and with 501c3 tax status to do educational outreach."

"Yep," agreed Mac. "We batted around some names. 'Naval Intelligence League' sounded too much like the other one. 'Naval Intelligence Retirees' sounded too geriatric."

Pete smiled. "We thought about Naval Maritime Intelligence Association, but that would have sounded like an enema."

"We wound up calling it NIP, even though we were afraid they would call us little NIPers." "Which they did," I said.

"The Foundation came about in a curious way. The Congressman from Virginia Beach got funding for a new schoolhouse at Dam Neck, since the USAF was going to close Lowry Air Force Base where we had our Air Intelligence School."

"That is the Navy-Marine Corps Intelligence Training Center, right? My younger boy may go there if he gets his commission this winter."

Mac nodded. "Interesting story, that is. They decided to name the building after Eddie Layton, the Fleet Intelligence officer for Chester Nimitz in the Pacific. Well, that ticked off Rufus Taylor's wife Karin, and she wasn't mollified by the fact that they named the auditorium after Rufus, who was the first designated intelligence specialist to make a three-star rank. She wanted to ensure that her husband was properly honored. She wanted to give NMITC skipper Bob Trafton $10,000 dollars to endow a fund for awards to the top graduate of the Basic Intelligence Course in Rufus Taylor's name."

"Bob went to Bill Studeman to ask if that was appropriate, and the legal opinion was that the active Navy couldn't take the money direct. We found a corporate lawyer in Richmond, and he set up the Foundation for us, and Shap brought an advisory panel of four-star officers to give it some kick."

"And now, 25 years into it, we are consolidating NIP and NIF," I said.

"Legal opinions change," Mac laughed. "Just think how you guys will have to adapt the organization to the new structure of Naval Intelligence, as it gets folded into Communications, Meteorology, Cryptology and Public Affairs. The new Information Dominance Corps reflects the way things were before World War Two, when the Office of Naval Intelligence got into it with War Plans, and the Radio Wars started in the Navy."

"You were there for all that," I said. "That might be worth a look back for the 25th Anniversary issue."

"It might," he said. "I was just a fresh-caught deck officer in 1941, but I can tell you quite a story. In the meantime, I need to get back to the Madison and hear the President explain things from the Oval Office."

"I'd be interested in an explanation for how this world came to be, Sir," I said, gathering up my notes. "Maybe we can get together at Willow again sometime soon."

I waved to Peter for the check, and reached for my wallet. That is the only part about Willow, but the staff treats me well, since I like to tip 100%. At least while I can.

SHOWERS, FOLLOWED BY SHOWERS

• • •

Mac chats with HRH Elizabeth-with-a-Z, 2005, at the dedication
of the WWII monument on the National Mall.

"What do you want me to do? Win World War Two again?"
–Shower's Family Reunion

Quote of the Week, 2012

It had been a marvelous weekend, filled with camaraderie and family, and it was an honor to be part of the gathering of the far-flung Showers clan and assorted admirers and friends as they celebrated the life of our friend, mentor and guide.

The night before had featured the money quote of a Showers Beach Week on North Carolina's Outer Banks—the sort of ultimate come-back to trump the contentions of a younger and sublimely confident generation to the revealed and settled wisdom of another.

The Showers family had secured the Williamsburg Room on the second floor of the luxurious new Arlington clubhouse of the Army-Navy Country Club or the memorial event. The golf season was just starting in earnest, late on a Sunday afternoon, golfers coming off the course and the place coming alive.

The old clubhouse, the one whose central core went back to the founding days of 1924, is fully gone, and the new contour of the ridgeline now resembles what it must have looked like then the Civil War Arlington Line of defenses snaked across the highlands. The outlines of the earthworks of Fort Richardson that now nestle the ninth green on the clubhouse approach are much more dramatic.

The property was part of the Nauk Neighborhood, whose immediate progenitor had been a Freedman's Village established at the end of the war. It is still true that the defining event in the life of the Boomers is World War Two. When we say "the War," or "after the War," as a marking point for some event in the social life of the nation, that is the defining moment.

But not here. The War still will always refer to the years that the Union Army surged across the River and constructed an astonishing ring of forts, protected firing positions, sunken roads, and all of them bristled with guns from the Arlington highlands.

MAC'S SISTER, IN FROM IOWA AND LOOKING SPIRITED AT 96 YEARS YOUNG.

Anyway, nice view from the new clubhouse, where the gathering of the Clan was just about complete. Mac's big sister was there, alert and vibrant with the same spirit of life that Mac always had. She was almost a dead ringer for her little brother, and with a walker just like his, and the same bright eyes that have seen nearly a century of life in These United States.

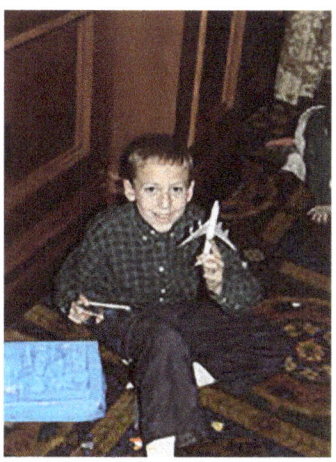

Mac's kids were there, and their kids, and a few of yet another generation crawling about with tiny trucks and jets, and some new arrivals still clinging to their mothers. Truly a multiple generational celebration of Mac's life and pretty damned impressive.

Son-in-law Tom brought in an extraordinary case of fine wine that had been a favorite of Mac's, whose traditional wine of choice had been Gallo Hearty Burgundy, a vintage I recall vaguely from Mom and Dad's adventures in fine wine, and which I despise to this day.

TWO OF THE MOST BEAUTIFUL GRANDDAUGHTERS EVER
SHARE MEMORIES OF THE MAN AND HIS TIMES.

Mac's tastes evolved over time, and if he did not have this precise vintage before, it was from a winery of which he would have thoroughly approved.

So, Saturday at Willow as the critical mass of family increased. An afternoon in the Gardens of Stone to survey the place of eternal rest for our pal, and his reunion with his beloved Billie. Now, this afternoon is the main event.

MAC AND HIS BELOVED BILLIE PREPARE FOR THE BIG TRIP.

Showers this morning, the meteorological version, and the weather guessers are saying things will dry up in time to accommodate the scheduling of the final military ceremony for Mac Showers this afternoon.

SCRIMSHAW

• • •

E ddie Layton's Captain picture. He appears this morning because of something special that happened last night. Bear with me—and photo thanks to Bill.

All right—this is one of those mornings. It is better than yesterday. That one started with Don-the-Builder's guys Tom and Jose showing up with the new back door at nine sharp at Refuge Farm, me on a conference call, and spiraled downhill or uphill from there.

There were such a dizzying number of things to consider: Burma and Gaza, one OK and the other not; Maryland and Rutgers joining the Big Ten, bringing the count to 14 teams, which doesn't make any sense, but what the hell. Maybe they will add another six teams and we can have the Big Ten Squared.

I was thinking about Mac as I drove back up north, crunching out of the driveway with two large men ripping the door frame out of the back of the house, and as far as I know, intending to seal it up again before they left. Michigan played Iowa, always a tough game regardless of who was having a decent season, and that was the only time that Mac took a certain pleasure in poking me. He was the complete gentleman, of course, but when his Hawkeyes kicked our butts, he was not shy about letting me know.

THE SELECTION OF WINE WAS FINE. LIQUOR MIGHT BE QUICKER, BUT THIS WAS VERY NICE INDEED. PHOTO: SOCOTRA

Anyway, for about the hundredth time I thought I needed to call him up, and the realization that I cannot still leaves me disoriented.

Mac's family was in town to do the estate sale for the unit he had at The Madison, and I got a kind invitation to go to dinner with them at Willow. I wandered over there after work, where things were at high PRF—pulse repetition frequency—for the big contract kick-off and the uncertainty that is swirling around the Government customer who does not have any more of a clue than the rest of us what is going to happen come this January.

I am sure everything is going to be fine, you know? I was working with Boats on further analysis of the oil-and-gas boom that is going to make everything fine, and will be quite useful as the global temperature cools over the next few decades. It will nice to be warm.

But anyway, Willow was welcoming in the growing dusk and the wine and stories flowed as they always do when Mac's family is together.

The table buzzed with conversation and laughter as always. Mac's daughter was having the same disassociation issue that I was— thinking I ought to talk to him about something or other, and her habit in that is life-long. Clearing things out from Mac's apartment and the attic at the house surfaced some interesting things. Between the salad and the entrée, she pulled out an ivory-colored box and handed it to me. There was an object of similarly colored material and a note yellowed with time along with a card.

Reading her note aloud made me blush, and I am not going to inflict it on you. But the square type-written provenance of the object brings us around to Eddie Layton, Mac's boss at PACFLT HQ during the war, and one of those historical figures who actually was in a position to change the world, and did.

Here is what his wife Miriam wrote to Mac and his wife Billie long ago, after Eddie passed on and Mac managed to get the Distinguished Service Medal presented to Joe Rochefort, the man responsible for providing the intelligence that stopped the Japanese cold at Midway:

Dear Mac and Billie –

My pleasure and gratitude to you for making it possible for me to be present at the Rochefort ceremony overflows! It was such a great thing to have accomplished.

For all your wonderful help and support I've looked around for something of Edwin's to send you. I know he would applaud this offering, I want it to represent my gratitude and affection for all you have done for me.

A bit of background—the first year Jack Kennedy was in the White House the media announced that Jackie was giving her husband a piece of scrimshaw for Christmas. Not to be outdone, I wrote to a jeweler in New Bedford who made scrimshaw about sending cufflinks for Edwin's Christmas, He sent me two pairs on approval. Edwin was so delighted with them that we kept them both.

Several years later when we went to New England, we visited the whaling museum in New Bedford where he bought several rough ugly

looking whale's teeth, learned the technique from the Jeweler and a
new hobby was launched,
He polished the tooth, drew his picture on thin paper and using a
dentist's scribe incised his design on the tooth and rubbed India Ink
into the scratch marks. I feel it would please Edwin to know you have
one of the four teeth he decorated with scrimshaw. He had great respect
and affection for you too. I enjoyed seeing you and Billie so much.
My only regret was that the time was so brief. Thank you so much
for your support and helpfulness over the last two years.
With love,
Miriam Layton
June 8, 1986

Mac typed the following note across the bottom of Miriam's letter:

> This letter accompanied the Layton-made
> scrimshaw received on 11 June 1986.

I was stunned. I was stunned. And that is how Eddie Layton, the fiery intelligence officer that Mac had described to me so many times, had decorated the small object from the sea. I remembered the story he told about the end of the Pacific War—the Japanese had given up, and the victorious Allies were converging on the Sagamai-wan to cement the victory with the signing of the treaty.

Eddie was there, along with the others who had guided the long struggle from Pearl Harbor to Tokyo Bay—and all the way to Arlington National Cemetery.

We were at Willow, of course, and I don't need to remember what Mac said, since I still have the notes on my collection of bar napkins. "Eddie was playing acey-duecy in the wardroom of *South Dakota* when Admiral Richmond Kelly Turner strode into the space. 'Terrible' Turner was in a state of high excitement. He was another of Admiral Nimitz' personal guests. He executed all the amphibious landings in the Pacific, from Guadalcanal to Iwo Jima, after all. But he was a son-of-a-bitch to his staff and liked the bottle. Eddie told me he was pretty fired up that evening. He started to shoot his mouth off and the wardroom hushed at the sound of the drunk four-star's booming voice."

"Terrible" Turner, a hero and mule-headed naval officer.

"What was he saying?" I asked.

"He was going off on Admiral Kimmel, of all things, the guy who was left holding the bag for the disaster that Terrible Turner caused by not passing critical radio intelligence to the commanders in Hawaii. According to Eddie, Turner was saying something to the effect that "Goddamned Kimmel had all the information and he didn't do anything about it. The court of inquiry said so, and they ought to hang him up higher than a kite!"

"But it was Turner himself who did not allow the critical Bomb Plot messages go to Pearl Harbor!" I exclaimed. "He must have known that. The court of inquiry was a white-wash to scape-goat Kimmel."

"You bet. Eddie sat there, stunned at what he was hearing. He had been there at the beginning before the attack, and then here at the end, the architect of the disaster was shouting that Kimmel ought to be hung up by his fingernails."

"I guess you can't do anything against four stars," I said thoughtfully, trying to imagine the scene in the wardroom of a big gray boat.

"Well, Eddie was pretty fired up, too. He corrected Turner in mid-rant. He told the Admiral that he had been there as Kimmel's intelligence officer, and he had been there in person."

"So what happened?"

"Eddie said the Admiral charged across the deck and grabbed him by the throat. Eddie was putting his dukes up to pummel the Admiral when the skipper of *South Dakota*, Emmet Forrestal, got in between them and broke it up."

I looked at Mac with amazement. The idea of decking a four-star Admiral made me admire Eddie Layton even more.

Looking over at the white tooth on my table, I have a little piece of something very special, created by the same hands that almost duked out a four-star hero. What an amazing day.

MEMORIAL

• • •

MAC SHOWERS VISITS HONOLULU FOR THE 71ST ANNIVERSARY
OF THE BATTLE OF MIDWAY IN 2012.

I t was 2012. Halloween was the day before and All Souls Day
for the memorial service. It is an appropriate one for Mac, and
held at the Faith Lutheran Church just east of my place at Big Pink
on Route 50.

I was really disoriented—Admiral Paul stopped by Willow yesterday afternoon at cocktail hour and I wasn't there yet, hung up

on some issue at work. Old Jim called me on the cell to summon me from my desk at the office across the street from the bar to chat at The Amen Corner.

The Macaroon Lady flew in from California to attend the service, and the regulars at the Willow bar entertained her. Then Mac's whole clan arrived, some in costume for the occasion, and we had plenty of wine and a fine meal in the dining room to commemorate the occasion. I clomped home later than usual and grateful to have power and now the Internet restored after one of Autumn's storms roared through.

I had to scour the place to find the bits of cloth and metal that will complete a Service Dress Blue uniform. It could be the last time I wear it, and appropriately, it is the one that my Dad saved from his service in WWII.

I jotted down some words for the ceremony. Here they are. It is not enough and it is not particularly eloquent. But it is certainly how we all feel:

"*It is a signal honor to be asked to talk a bit about the military career of our pal Mac. I still cannot quite believe that he is gone. When you knew the Admiral, in time you just came to assume that he was eternal.*

It was important for Mac to help people understand things. He certainly helped me. We starting talking several years ago. I enjoyed his company, and he enjoyed the beer, back when he could still get his doctor to agree to let him drink one or two. We spent hours and hours talking about his life and times.

As you might imagine, he had his stories down pretty well after seven decades, and I enjoyed mixing things up. Sometimes we would start on one thing and wind up somewhere else—like life in Depression-era Iowa, with the banks closed and only a barter economy enabling people to get by.

That he had to bring his recruiting officer to meet his mother Hedwig in order to get her permission for him to enlist.

I think you know the amazing events in which he played a part. The three that everyone knows are the first even fight in the Pacific at the Battle of the Coral Sea. Then the triumph of the code-breaker's art in the epic Battle off Midway Atoll. Before that encounter, the Americans never won a battle against the Imperial Japanese, and after it, they never lost one.

You all have heard the story about how Station HYPO identified Admiral Yamamoto's flight itinerary and enabled Air Corps P-38s to intercept and shoot him down in the Solomons in April of 1943. It is still controversial, though it was not to Admiral Nimitz at the time. "Kill the S.O.B.," he said.

On the way through these famous tales, I found some things that just plain amazed me.

Mac was a Deck Officer—what we would know now as a Surface Warfare officer—though he never served on a ship. Big Navy had no idea what Station HYPO was up to, and periodically they would ask Mac to go to sea. As part of that, legendary submarine skipper "Mush" Morton asked him to go on a war cruise on USS Wahoo as a sort of orientation to the art of submarine warfare.

Mac mentioned that to me one evening as I was settling up my tab at Willow, and I casually asked him what happened.

"Jasper Holmes would not let me go. The guy I was supposed to relieve got off the sub, and they had to go one officer short in the wardroom. Good thing. They never came back."

Which brings me around to the notion of fate. We agreed that much of his legendary career—at least at the beginning—was dumb luck. Half his class at Investigation School—the top of the alphabet—was ordered to Corregidor. In other words, from graduation to prisoners of war, just like that. Mac went to Pearl, just eight weeks after the attack, and the great ships still sunk in the mud of the harbor.

Dumb luck that the Officer in Charge of the Counter-intelligence office in Honolulu was unimpressed with Mac's experience, and sent him off to that obscure billet working for Joe Rochefort.

By 1945 he was with Fleet Admiral Nimitz, closing the ring on the Empire of Japan in Operation Starvation, slipping target nominations on the sly to General Curtis "Iron Pants" LeMay. Walking around Yokosuka, Japan, five days after the surrender on a "courier" mission that his boss Eddie Layton arranged so he could see it. That is where that giant Japanese flag that is out at ONI came from, traded for a bottle of Three Feathers Whiskey to a young Marine guarding the last floating Japanese Battleship, IJN Nagato.

Amazing at every turn. After the war, bumping by chance into Admiral Forrest Sherman in the halls of Main Navy the very week of the transfer board that would establish the new structure of Naval Intelligence, and becoming an intelligence officer.

Convincing Marshall Tito's people that he should have lunch with the Yugoslav leader in The White Palace.

Turning a so-so assignment at the Intelligence School at Anacostia into the first genuine OPINTEL course to pass along the techniques that won a global war.

Suggesting that his colleague Rufus Taylor to transfer to intelligence at Arlington Hall Station. Ruf became the first intelligence professional to become the DNI, and the first to wear the three stars of a Vice Admiral.

At First Fleet, deciding to provide target materials to Navy pilots assigned to carry atomic weapons to use against the Soviet Union.

Returning to Pacific Fleet Headquarters as the Fleet Intelligence Officer—the same job in which Eddie Layton served in World War 2—to confront the conflict in SE Asia. Then Washington again, and Purple Dragon and the Pueblo Damage Assessment, and that big deal with Howard Hughes we are still not supposed to talk about.

And that is where we wander into things he preferred not to speak about on the record, but for which he was recruited to CIA by the legendary Bronson Tweedy, turning in his letter to retire on December 31, 1971.

I have paper napkins and notebooks with all of it, and since Mac reviewed them all, I feel that the unofficial and un-footnoted story will be true and accurate.

When he was done with government service in 1983, he started a third career. His experience with the cruelest disease—Alzheimer's—helped me through the decline and loss of my Mom and Dad.

I am going to miss him a lot. When I was typing this, I had the weirdest sense that I needed to call him up and ask him a question.

At the end of the day, what Mac Showers did was make a career out of helping people to understand things. What a man he was, and what a legacy. We were all honored to have shared the planet with him, and I am absolutely confident that I will see him again for liberty on the other shore.

So long, Shipmate."

OBITUARY

• • •

Naturally, Mac was always prepared. He decided to draft his death notice and have it ready for the family when it was needed. We did some of the editing near the Amen Corner at the Willow Bar. It was only appropriate:

Donald McCollister Showers was born Aug 25, 1919, in Iowa City, Iowa, the son of Charles N. Showers (1888–1973) and Hedwig Marie *Potratz* (1889–1980). He preferred to be called "Mac," an abbreviation of his middle name, McCollister, his paternal grandmother's maiden name.

Mac graduated from the State University of Iowa with a B.A. and certificate in journalism. His interest in the Navy led him to enroll in the Navy Reserve's fast-track V-7 program and he attended the Northwestern University Naval Reserve Midshipman's School in Chicago.

Mac was commissioned an ensign on Sep 12, 1941. Trained in counterintelligence (CI), he reported for duty at the District Intelligence Office (DIO) in Honolulu, Hawaii. But DIO needed experienced investigators, so in February

1942 he was reassigned to the Combat Intelligence Unit, Station HYPO, Pearl Harbor, Hawaii commanded by CDR Joseph Rochefort (1900–1976).

Although he had no background in intelligence or code breaking, Mac quickly integrated himself to the mission of HYPO, which was to break the Japanese Naval code, JN-25. This was critical to provide ADM Chester Nimitz (1885–1966) with the intelligence he needed to best situate his forces for coming battles. Because of the intelligence that HYPO provided, Nimitz situated his forces in the best possible position to achieve surprise on the Japanese Navy at the Battle of Midway in June 1942.

When Rochefort left HYPO in the summer of 1942, Mac remained, serving under Rochefort's relief, CDR William Goggins (1898–1985). Mac played an important role in the 1943 shootdown of the airplane carrying Japanese Admiral Isoroku Yamamoto. When Yamamoto's itinerary was intercepted, Mac analyzed it to determine if it was a plausible schedule or not. It was, and his analysis led, in part, to the decision to target Yamamoto.

In 1944, Mac was selected to serve as deputy to CAPT Edwin Layton (1903–1984), Admiral Nimitz's intelligence officer, to establish a fleet combat intelligence center (Advance Intelligence Center) on Guam. While at Guam AIC, he gave daily intelligence briefs and provided intelligence to the planners of the invasion of Japan.

In Oct 1946, Mac became one of the first naval officers to be designated an intelligence specialist.

On Jun 12, 1948, he married Sarah Vivian "Billie" *Gilliland* (1923–2002). They had a daughter and two sons.

Mac was selected for rear admiral in 1965, thus becoming the only member of HYPO to attain flag rank. During the Vietnam War, he was Fleet Intelligence Officer to the Commander in Chief of the Pacific Fleet, an assignment that CAPT Layton had previously held.

His last assignment was as Chief of Staff at the Defense Intelligence Agency (DIA). While at DIA, Admiral Showers was charged with heading an effort, called PURPLE DRAGON, to find and correct breaches of security which had become known to the North Vietnamese during the Vietnam War. Its findings led to greater combat effectiveness and saved lives in Vietnam, and its lessons were applied by the military to its operations in general.

Admiral Showers retired from the Navy on Dec 31, 1971. During his career, he had been awarded the Navy Distinguished Service Medal, the Legion of Merit and the Bronze Star.

Three days after retiring from the Navy, he embarked on a civilian career with the Central Intelligence Agency. In the early 1980s, he served as special assistant to CIA Director William Casey (1913–1987).

After 12 years with the CIA, he retired a second time in

order to become a full-time caregiver to his wife, Billie, who was diagnosed with Alzheimer's disease. Billie died in 2002.

After his retirement from CIA, Admiral Showers continued to lead an active life. He was helped to fundraise to find a cure for Alzheimer's. He traveled and spoke to groups, both civilian and military, about his experience at HYPO.

Admiral Showers was central to the effort to honor CAPT Rochefort's efforts at HYPO. Admiral Nimitz had nominated CAPT Rochefort for the Navy Distinguished Service Medal (DSM), but it had been disapproved by the Chief of Naval Operations, ADM Ernest King (1878–1956). Although CAPT Rochefort had died in 1976, Admiral Showers shared the opinion of other surviving members of HYPO that CAPT Rochefort was deserving of the DSM.

In 1981 Admiral Showers wrote and submitted the paperwork to renominate CAPT Rochefort for the DSM. After some disappointment and an initial lack of success, the Secretary of the Navy eventually approved the award in 1986. Admiral Showers was able to arrange for the DSM to be presented to the Rochefort family by President Reagan. The presentation ceremony took place on May 30, 1986 at the White House with Admiral Showers in attendance.

Although he never wrote a memoir of his own, Admiral Showers assisted others in sharing their memories of their service in the war. He assisted his HYPO shipmate, CAPT Jasper Holmes (1900–1986), in the writing of his 1979 book, *Double-Edged Secrets: U.S. Naval Intelligence Operations in the Pacific*

During World War II. He helped Edwin Layton (who also retired a rear admiral) when Admiral Layton wrote his 1985 memoir, *And I Was There: Pearl Harbor and Midway—Breaking the Secrets.* In 2011, he wrote the foreword to Elliott Carlson's biography of CAPT Rochefort, *Joe Rochefort's War: The Odyssey of the Code Breaker who Outwitted Yamamoto at Midway.*

The Navy Marine Corps Intelligence Training Center in Dam Neck, Virginia, presents the "Rear Admiral Donald M. Showers Award" to the Honor Graduate of its Naval Intelligence Officers Basic Course.

In 2008 Admiral Showers was inducted into the National Security Agency's Cryptologic Hall of Honor, an honor previously extended to his shipmates at HYPO, CAPT Rochefort, and CAPT Thomas Dyer (1902–1985).

Admiral Showers died Oct 19, 2012, at the Virginia Hospital Center in Arlington, Virginia with his two sons and daughter at his side. He was 93.

AUTHOR BIO

Vic Socotra

Vɪᴄ Sᴏᴄᴏᴛʀᴀ was the Air Wing SIX Inteligence Officer on the 1989–1990 cruise of USS *Forrestal* to the Mediterranean Sea. It was to be another in the ninety-odd such aircraft carrier deployments to that body of water, and was scheduled to accomplish the deterrance and contingency missions that had been common since the close of the second world war and the cold war. Something unusual occurred during this deployment that made it unique. While deployed, the Berlin Wall fell, President Bush flew to Malta to meet Soviet Leader Gorbechev, and the 45-year Cold War essentially ended.